ECOLOGICAL GENETICS

ECOLOGICAL GENETICS

Edited by Leslie A. Real

PRINCETON UNIVERSITY PRESS   PRINCETON, NEW JERSEY

Library of Congress Cataloging-in-Publication Data

Ecological genetics / edited by Leslie A. Real.
p.   cm.
Includes bibliographical references (p.) and index.
ISBN 0-691-03241-6 (cl) — ISBN 0-691-00066-2 (pb)
1. Ecological genetics.   2. Evolution (Biology)
3. Population genetics.   I. Real, Leslie.
QH456.E26   1993
575.1—dc20   93-4762

This book has been composed in Adobe Sabon

Princeton University Press books are printed
on acid-free paper and meet the guidelines
for permanence and durability of the Committee
on Production Guidelines for Book Longevity
of the Council on Library Resources

Printed in the United States of America

10   9   8   7   6   5   4   3   2   1

(Pbk.)
10   9   8   7   6   5   4   3   2   1

Contents

List of Contributors

JANIS ANTONOVICS
Department of Botany
Duke University
Durham, NC 27706

ANDREW JAROSZ
Department of Botany and Plant
Pathology
Michigan State University
East Lansing, MI 48824

MICHAEL LYNCH
Department of Biology
University of Oregon
Eugene, OR 97403

LESLIE A. REAL
Department of Biology
Jordan Hall
Indiana University
Bloomington, IN 47405

MONTGOMERY SLATKIN
Department of Integrative Biology
University of California
Berkeley, CA 94720

KEN SPITZE
Department of Biology
University of Oregon
Eugene, OR 97403

DON STRATTON
Department of Ecology and Evolution
Princeton University
Princeton, NJ 08544-1003

PETER THRALL
Department of Botany
Duke University
Durham, NC 27706

JOSEPH TRAVIS
Department of Biological Science
B-142
Florida State University
Tallahassee, FL 32306-2043

SARA VIA
Department of Entomology and
Section of Ecology and Systematics
Comstock Hall
Cornell University
Ithaca, NY 14853

Introduction

Current Directions in Ecological Genetics

Leslie A. Real

IN HIS CLASSIC *Ecological Genetics,* published close to thirty years ago, E. B. Ford helped define the field by characterizing ecological genetics as a unique area of scientific inquiry combining ecological fieldwork and laboratory genetics. Ford's book summarized almost three decades of his own and others' research in the new field of ecological genetics and was instrumental in developing the modern evolutionary synthesis. Ford's book emphasized understanding the significance of genetic mechanisms in the evolution of population phenotypes within the present time and within the natural conditions of the organism. Writing three decades ago, Ford asserts:

> The term "ecological genetics," which describes the technique of combined field and laboratory work outlined here, has recently come into general use. I have, however, for many years employed it in lectures and scientific discussions, in which it has proved self-explanatory. Ecology, which denotes the inter-relation of organisms with one another and with the environment in which they live, may be regarded as scientific natural history. Consequently, ecological genetics deals with the adjustments and adaptations of wild populations to their environment. It is thus, . . . essentially evolutionary in outlook. Indeed it supplies the means, and the only direct means, of investigating the actual process of evolution taking place at the present time.

The definition of ecological genetics provided by Ford characterizes contemporary approaches to studying evolution in natural populations. What then constitutes the new excitement in the field, and what new advances are occurring half a century after the great era of ecological genetics represented by the Modern Synthesis (Mayr and Provine 1980)?

During the spring of 1991, five distinguished ecological geneticists were invited to the University of North Carolina to address these questions. Each speaker prepared two talks: one on a specific research topic, and a second more general and synthetic overview of an important emerging research area in contemporary ecological genetics. The ten articles represented here are based on these ten addresses.

I have arranged the book by author rather than by topic for two reasons. First, each pair of papers represents a particular "vision" as to the direction of future research. The authors were encouraged to be quite personal in their approach and perspective. Second, while authors were invited to cover some specific topic—for example, phenotypic plasticity, gene flow, or life-history evolution—most often authors addressed issues that cut across specific topics. Important issues and areas appear in every one of the discussions. In reviewing all of the chapters, I was amazed to see the degree to which topics cannot be neatly subdivided as they have been in the past. It would be impossible to arrange this book by chapters on, for example, mutation, migration, selection, mating system, and so forth in a manner analogous to the most popular population genetics texts of the previous generation (e.g., Wallace 1968; Dobzhansky 1970; Grant 1963). All of the chapters share a common "tone" that distinguishes them from the more classical approaches of the founders of ecological genetics. What exactly constitutes this new tone, and what are the common features shared by the chapters?

The classical ecological geneticists were principally concerned with documenting and measuring the magnitude of selection in natural systems. E. B. Ford cites as the first case study in ecological genetics an analysis by Gerould (1921) of the selective elimination of a recessive gene in the butterfly *Colias philodice* through visual predation. The documentation of natural selection in the wild remains an important area of research (recently summarized by Endler 1986). However, the classic approach generally concentrated on selection within single populations in which the exact size of the population was not an essential feature, and on aspects of the phenotype (e.g., visible polymorphism) under simple genic control. The contemporary approach seems, on the other hand, to emphasize the importance of population size and structure, the interaction among populations through migration and dispersal (i.e., the "metapopulation" structure), the interaction between local selection and genetic drift, and an expansion of the phenotype to include quantitative (as well as qualitative) characters.

The new approach appears to have been fueled by several methodological and conceptual advances over the last two decades. Two methodological developments stand out. First, molecular techniques for decomposing the organism's genotype (e.g., electrophoresis, amino acid and DNA sequencing) have become readily available. Second, we have extended the theory and application of the statistical analysis of quantitative characters and the partitioning of variance as a predictor of selection response. These two methodologies, ironically, stand at opposite poles to each other. One emphasizes genetic analysis at the extremely small scale of even single nucleotides, while the other averages effects over multiple loci without particular regard to molecular basis. The two methodologies are used to address and answer different sets of questions, but both have proven invaluable.

Molecular techniques are widely used by the authors in this book. Lynch (chapter 6), for example, uses both isozyme and restriction fragment length polymorphisms (RFLPs) to assess the level of heterozygosity as an indicator of mode of reproduction, isolation, and subdivision among natural populations of *Daphnia*. Slatkin (chapters 1 and 2) uses nucleotide sequence data to determine rates of gene flow among populations undergoing different degrees and modes of isolation and population subdivision and "coalescence" events in cladistic analyses.

Quantitative genetic analysis is used by Via (chapters 3 and 4) and Travis (chapters 9 and 10) to determine the evolution of life-history characteristics, evaluate the extent of local adaptation, and directly assess selection on fitness. Lynch (chapter 5) constructs models of the evolution of quantitative traits under a variety of ecological conditions representing different types of population subdivision and isolation.

The methodological advances alone cannot account for the great excitement that now flourishes in evolution and ecological genetics. These two technical advances have been linked to several important conceptual advances, the most important of which seem to be the emergence of a rich theory for the evolution of neutral characters, the operation of ecological processes coupled to evolutionary processes at the metapopulation level, and a focusing of attention on life-history phenomena relating genetic processes and population dynamics.

The neutral theory, largely developed through the pioneering efforts of Motoo Kimura (1969, 1983) and King and Jukes (1969), suggest that amino acid substitutions and evolution at the molecular level are primarily due to the action of genetic drift rather than to natural selection. Few amino acid substitutions should have selective consequences. The mathematical theory derived from assumptions of random genetic drift allow for specific predictions concerning rates of allele frequency change both within and among populations. Rates of allelic change, under the assumption of neutrality, are a function of population characteristics such as size, population structure, and isolation and migration, as well as genic considerations such as mutation rate. By determining frequencies of molecular variations, one can assess aspects of population structure. The introduction of molecular techniques into population genetics permits the ecological geneticist to monitor population structure and process in a manner more direct than possible with any previous technique.

The application of molecular techniques coupled with the predictions of the neutral theory is used by Slatkin (chapter 1) to examine various models of population structure, to measure gene flow, and to characterize modes of population isolation. Similar techniques are employed by Lynch in his papers. While Slatkin's analyses, and the efforts of most molecular evolutionists, are directed at qualitative genetic characters (e.g., nucleotide sequences), Lynch (chapter 5) goes further by extending the neutral theory to quantitative phenotypic characters. He shows that much of the predictive value of the neutral theory, and our ability to determine population structure by adopting the assumptions of the neutral theory, hold for quantitative characters. It should be noted, however, that the determination of population structure using neutral theory often depends on the assumption that very little or no selection is occurring at the molecular level, and that mutations accumulate at a measurable rate (the "molecular clock"). These two assumptions are under intensive scrutiny at the moment, and there is considerable evidence suggesting that the assumptions of the neutral theory may not hold in many cases (see Gillespie 1991 for a recent summary of the evidence for and against the neutral theory). Nonetheless, neutral theory can be quite useful in detecting the operation of selection by indicating the direction of deviation from expectation. For example, Lynch (chapter 6) postulates that fluctuating selection within populations inflates genetic variation above that expected from neutral theory.

The results of molecular analyses suggest that population size and migration dynamics among populations is much more important than previously thought. The only way to truly understand evolution in natural systems is through the consideration of the metapopulation, that is, the set of interconnected subpopulations that differ in size and are subject to different degrees of migration, gene flow, and local adaptation.

The metapopulation approach is adopted by Antonovics in his treatment of plant-pathogen dynamics and population genetics. Antonovics (chapter 7) first examines the interplay between gene frequency change and absolute fitness in the *Silene alba–Ustilago violacea* plant-pathogen system. He explores the consequences of both frequency-dependent and density-dependent modes of disease transmission within a given population. In many cases where the ecologically important characteristics of the population are held constant (e.g., numbers of infected individuals versus healthy individuals), future trajectories of the populations may differ when the underlying genetic composition of populations differs. Antonovics and his colleagues (chapter 8) then proceed to examine the consequences of different magnitudes of migration among populations for disease dynamics. One conclusion is that only small amounts of migration are required to disrupt local population dynamics governed primarily through demographic and genetic stochasticities. The metapopulation approach explicitly acknowledges the nonequilibrium nature of population processes. New populations are founded, infections are introduced through migration events, and local populations go extinct. The emerging pattern becomes apparent only over larger geographic and spatial scales.

One of the more interesting and important observations that follows from the *Silene-Ustilago* study is the importance of underlying species genetics in determining the outcome of species interactions. Ecological genetics has principally focused on single-species systems. Yet, as Antonovics and his colleagues demonstrate, evolutionary change will often be most influenced by the interaction among species where the strength and nature of the interaction will be determined, in large part, by genetic factors. The implications of species interactions for understanding evolutionary processes has led Antonovics (chapter 8) to call for a "community genetics" that will extend traditional approaches beyond a focus on single species.

An emphasis on the relation among genetics, fitness measured in absolute terms (e.g., numbers of offspring, age of first reproduction, etc.), and population dynamics naturally leads to an increased need for understanding the evolutionary ecology of life histories. Both Travis and Via explore the ecological genetics of important life-history phenomena and consider how to measure lifetime fitness and the importance of placing life-history characteristics (e.g., fecundity) in a demographic context. Both authors argue that there are often strong genotype-environment interactions in life-history traits and illustrate the need for analyses across populations subject to different environmental influences. The genetic covariance among traits can vary across environments, suggesting that life-history phenomena are quite plastic within and among populations. Travis and Via provide splendid examples of the means necessary to disentangle the complicated forces that generate the patterns of age-specific reproduction and survival which ultimately lead to population growth and fitness.

A new ecological genetics that focuses on large-scale geographic variation in demographic and genetic dynamics among small, partially isolated, and potentially locally adapted populations has obvious implications for conservation and natural-resource management. As habitats are fragmented through human intervention, populations that at one time were large and interconnected are now becoming disconnected and locally rare. Habitat fragmentation makes migration critical for the reestablishment of local populations, and the concomitant gene flow may counter the detrimental effects of inbreeding depression in rare populations. An anonymous reviewer has commented:

Conservation considerations *force* us to examine interactions between population genetics and population size dynamics, gene flow, and migration or local replenishment. Geneticist interested in inbreeding depression and other aspects of genetics of rare species must consider fitness effects in absolute terms, and thus address reproductive rate and density-dependent regulation of population size and arguably age structure and other aspects of within-population demographics. Population ecologists working on small populations (often extremely small populations in conservation efforts) cannot ignore the potentially devastating effects of inbreeding depression and loss of genetic variation for resistance to pathogens.

Populations that are small and locally adapted may also suffer under increased gene flow through the disruption of locally adapted genotypes. Migrants differ in their genotypes and are not genetically equal in terms of their effects on local population growth and fitness. Recognizing the genetic diversity that exists among migrants seems essential when constructing conservation policies for reintroducing or augmenting declining populations of important species.

Conservation biology is not the only area where ecological genetics will be a valuable applied tool. The long-term health of ecological systems will depend upon the ability of organisms to adapt and respond to changes in their biotic and abiotic environment. The methods for studying adaptive evolution in contemporary populations provided by modern ecological genetics will prove equally important in assessing the ability of organisms to respond to both local and global aspects of environmental change, including response to anthropogenic stresses in disturbed ecosystems.

The papers presented in this volume are only a sample of the possible new directions for ecological genetics and represent only a handful of the creative scientists engaged in this important research. This book could have easily been two or three times its current size. Nonetheless, the topics and research programs presented are rich in new insights and perspectives and represent a broad spectrum of the current enthusiasm and creative energies of practicing evolutionary biologists. The ground-breaking work represented here will certainly occupy the minds and talents of the next generation of ecological geneticists.

References

Dobzhansky, Th. 1970. *Genetics of the Evolutionary Process*. Columbia University Press, New York.
Endler, J. A. 1986. *Natural Selection in the Wild*. Princeton University Press, Princeton, N.J.
Ford, E. B. 1964. *Ecological Genetics*. Methuen and Co., London.
Gerould, J. H. 1921. Blue-green caterpillars. *J. Exp. Zool.* 34:385–412.

Gillespie, J. H. 1991. *The Causes of Molecular Evolution.* Oxford University Press, Oxford.

Grant, V. 1963. *The Origin of Adaptation.* Columbia University Press, New York.

Kimura, M. 1969. The rate of molecular evolution considered from the standpoint of population genetics. *Proc. Natl. Acad. Sci. USA* 63: 1181–1188.

Kimura, M. 1983. *The Neutral Allele Theory of Molecular Evolution.* Cambridge University Press, Cambridge, U.K.

King, J. L., and T. H. Jukes. 1969. Non-Darwinian evolution. *Science* 164: 788–798.

Mayr, E., and W. B. Provine. 1980. *The Evolutionary Synthesis.* Harvard University Press, Cambridge, Mass.

Wallace, B. 1968. *Topics in Population Genetics.* W. W. Norton, New York.

ECOLOGICAL GENETICS

1

Gene Flow
and Population Structure

POPULATION structure consists of two distinct but interrelated parts: the demographic structure and the genetic structure. The demographic structure is determined by all the processes associated with birth, death, and dispersal, including the mating system and life history. The genetic structure is determined by the population structure, of course, but also by genetic processes such as selection, recombination, and mutation.

I make the distinction between these two parts of population structure because of the different tools needed to examine each part. The demographic structure can be studied by censusing and observation. From such studies come estimates of birth and death rates, population densities, dispersal distances, extinction and colonization rates, and the characteristics of the breeding system. No knowledge of the genetic basis of any of these characters is needed. To determine the genetic structure, it is necessary to understand the pattern of genetic variation in the species, which means assessing the genotypes of different individuals. In the past twenty-five years, developments in biochemical methodology have allowed the examination of many parts of the genome using protein electrophoresis, restriction endonucleases, and DNA sequencing. Despite the tremendous advances in methodology that have been made, however, only a minute fraction of the genome of any species has been examined and relatively few species have been studied in any detail. Additional information about the genetic structure of natural populations has been provided by the recent use of statistical methods from quantitative genetics to estimate heritabilities and genetic correlations of phenotypic characters.

In this chapter, I will discuss two aspects of population structure. The first is the relationship between the demographic and genetic structure of a population, with particular concern about what features of the genetic structure can be predicted from knowledge of the genetic structure, and the second is the extent to which information about the genetic structure allows us to draw conclusions about the demographic structure, particularly the levels and patterns of gene flow.

Sewall Wright on Population Structure

Of the three pioneers of population genetics theory—Fisher, Haldane, and Wright—Wright was by far the most concerned with population structure and its role in evolution. Haldane and Fisher focused much more on evolution in large randomly mating populations in which population structure was not an issue of importance. Much of Wright's theoretical work can be regarded as showing how the demographic structure of a population determines the genetic structure. Wright recognized the tremendous diversity of demographic structures that a species could have, and he tried whenever possible to reduce that complexity to a few simple quantities that would suffice to predict major features of genetic structure.

Wright's early work on determining the coefficient of inbreeding from a pedigree is an excellent example of Wright's methods. He showed how the effect of any pattern of ancestry on the correlation of genetic state of a locus could be summarized by a single quantity, the inbreeding coefficient, which could be determined by tracing paths of ancestry in a pedigree (summarized in Wright 1969, chap. 7). Wright's method is such a basic part of population genetics theory that it is easy to forget what an achievement it was. There was no reason to assume in advance that inbreeding in different pedigrees could be so easily characterized. Different kinds of inbreeding could result from different kinds of common ancestry. Wright showed that a single number was sufficient.

Wright's theory of effective population size is a result of the same type. He showed that complexities of a breeding system such as sex ratio or effects of temporal variation in population size can be summarized in a single number—the effective population size—that determines the overall strength of genetic drift in a population. In this case, later work by Kimura and Crow (1963), Ewens (1989), and others have shown that different definitions of effective population size are needed for different purposes.

Wright's analysis of geographically subdivided populations had the same goal. He showed that in many cases the effect of restricted dispersal could be described by a single quantity. In his "island model," in which each local population replaced a fraction m of its residents with individuals chosen at random from a large collection of local populations, he showed that the variance in allele frequency among islands is approximately

$$V_p = \frac{\bar{p}(1-\bar{p})}{1 + 4Nm},$$

where N is the effective population size of each local population, m is the immigration rate, and $\bar{p}$ is the average frequency of the allele among the immigrants (Wright 1931). The variance depends only on the combination of parameters, Nm, and not on each parameter separately.

Later, Wright considered the problem of isolation by distance (Wright 1943). He modeled the situation in which dispersal in species occupying a large geographic area was restricted. Wright attempted to compute the variance in allele frequencies among local areas and the relationship of inbreeding coefficients in areas of different size. He concluded that what he defined as the neighborhood size—the number of individuals in an area from which the parents of each individual were sampled—was equivalent to the effective population size of a single population and was sufficient to predict the extent of variation in allele frequencies among locations.

Other theoretical work on the problem of isolation by distance, particularly by Malécot (1968) and by Kimura and Weiss (1964), showed that Wright's (1943) analysis was incomplete. These authors showed that in fact the variance among allele frequencies in different areas cannot, as Wright had claimed, be computed only from the neighborhood size. For neutral alleles at a locus, the variance in allele frequency among location depends both on the neighborhood size and on the mutation rate at that locus. Furthermore, the decrease in covariance in allele frequencies with geographic distance depends on the mutation rate as well. It is, then, not possible to achieve Wright's goal of characterizing the effect of isolation by distance strictly in terms of the demographic structure of the population, at least by using genetic variances and covariances. The neighborhood size depends on dispersal distances and population densities, which are part of the demographic structure, but the mutation rate depends on genetic processes that will vary from locus to locus. I will discuss later how Wright's goal can be achieved by asking the question in a slightly different way than Wright did.

Gene Flow and Population Structure

One of the problems in the analysis of population structure is determining the amount of gene flow. Gene flow is a major component of population structure because it determines the extent to which each local population of a species is an independent evolutionary unit. If there is a large amount of gene flow among local populations, then the collection of populations evolves together; but if there is little gene flow each population evolves almost independently.

How much gene flow is needed to prevent independent evolution in different local populations depends on what other forces are at work. Wright's result for an island model tells us the balance achieved between gene flow and genetic drift. If Nm is much greater than 1, then gene flow overcomes the effects of drift and prevents local differentiation. If Nm is much less than 1, then drift acts almost independently in each population. More recent work (Takahata 1983) has shown that it takes a population a time τ, which is approximately the larger of $2N$ and $1/m$, to reach this equilibrium. Before τ the genetic composition will be determined primarily by the initial conditions, and after τ the composition will be determined by Nm.

Haldane (1930) considered the problem of the balance between gene flow and selection. He assumed that immigrants to an island carried an allele a that is deleterious on the island. If the selection coefficient in favor of the other allele, **A**, on the island is s and the immigration rate is m, then the equilibrium frequency of **A** on the island is $p = 1 - s/m$ if $s < m$ and 0 if $s > m$. The time it takes to approach this equilibrium is approximately $1/|s-m|$. This result has been extended by Nagylaki (1975) to more general selection models. Of course, selection that favors the same allele in different populations will generally prevent differentiation regardless of the amount of gene flow.

These theoretical results tell us that gene flow might prevent differentiation at some loci—those which are neutral or are weakly selected—but not at loci that are strongly selected. Furthermore, the time it takes to approach the equilibrium states of each locus depends on both the extent of gene flow, as measured by m, and the strength of selection, as measured by s. Strongly selected loci will reach their equilibrium frequencies much more rapidly than neutral or weakly selected loci.

Methods for Estimating Levels of Gene Flow

I have distinguished two classes of methods for estimating levels of gene flow in natural populations (Slatkin 1985a). "Direct" methods are those that depend on observations or experiments that measure the extent of dispersal. For example, mark-release-recapture studies can provide estimates of the mean distance between the point of release and the point of recapture. Estimates of dispersal distances can be converted into estimates of gene flow if it is assumed that dispersing individuals have the same chance to mate as do residents. It is also possible to follow the progress of distinctive alleles into a population not containing those alleles, thus showing that gene flow has occurred and learning about its properties (Bateman 1950; Handel 1982). A slightly different method is to use genetic methods to assess paternity and compute the distance between parents and their offspring (Ellstrand and Marshall 1985).

In contrast to direct methods, "indirect" methods do not depend on studies of dispersing individuals. Instead, they lead to estimates of the average level of gene flow from a mathematical model of the interaction of gene flow and other forces to predict how much gene flow must have been occurring in order for patterns observed in the data to be present. The most commonly used indirect method is based on Wright's F_{ST} statistic, which is a measure of the correlation between genes in a subpopulation relative to the entire population (Wright 1951). For two alleles at a locus, F_{ST} is approximately $V_p/\bar{p}(1-\bar{p})$, where V_p is the variance in allele frequency among subpopulations and $\bar{p}$ is the average frequency. Wright (1951) showed that in an island model at equilibrium,

$$F_{ST} \approx \frac{1}{1 + 4Nm}.$$

From this result an indirect estimate of Nm can be obtained by computing F_{ST} from allele frequency data and then finding Nm from

$$Nm \approx \frac{1}{4}\left(\frac{1}{F_{ST}} - 1\right). \tag{1.1}$$

Wright (1969) and others have used this method extensively to estimate Nm values in different species. There is no indirect method to estimate m separately, but that value can be obtained from Nm and from estimates of N obtained using census data.

Both direct and indirect estimates of levels of gene flow reveal different things about a species, and both methods have their strengths and weaknesses. Direct estimates indicate the details of dispersal, including the life-history stage during which dispersal occurs and possibly also the ecological conditions favorable to dispersal. Direct estimates have the disadvantage that they are based on studies that are necessarily limited in scale. It is very difficult to obtain information about occasional long-distance dispersers or to obtain information about dispersal under unusual environmental conditions. There is the additional problem that dispersing individuals may be unable to find a mate or to successfully raise their offspring. Indirect estimates, on the other hand, have the advantage that they incorporate the effects of all kinds of dispersal and effectively average over variation in dispersal through time. Indirect methods have the disadvantage that they depend on assumptions about processes affecting allele frequency, and those assumptions cannot be independently tested. For example, indirect estimates assume that a species is at a genetic and demographic equilibrium, something that may not be true.

Properties of F_{ST} as an Indirect Estimator of Nm

For an indirect estimator of the average of Nm to be useful, it should have three desirable properties. First, it should be relatively insensitive to forces other than gene flow and genetic drift. In particular, it should be insensitive to the mutation rate and the intensity of selection, if any, affecting a locus. If that were not the case, then estimates based on allele frequencies at different loci would not be expected to agree because it seems likely that mutation rates and selective regimes would differ among loci. Second, an indirect estimator of the average level of gene flow should not be sensitive to the actual pattern of gene flow in a population, although as discussed later, it is possible to develop indirect estimators that do indicate patterns of gene flow. Third, an indirect estimator should approach its equilibrium value relatively rapidly to ensure that estimates reflect ongoing gene flow as much as possible.

F_{ST} approaches its equilibrium value more quickly than some other measures of genetic variability, particularly the heterozygosity, although it still may approach its equilibrium value slowly in large populations. Takahata (1983) and Crow and Aoki (1984) showed that F_{ST} at equilibrium in the n-island model for a locus with a mutation rate μ to alleles not previously found in the population is approximately independent of the mutation rate as long as the mutation rate is much less then the migration rate, a condition that is biologically reasonable. Slatkin and Barton (1989) found that F_{ST} in an island model is not very sensitive to selection favoring one or two alleles in each population in an island model. There is one condition under which selection will affect the value of F_{ST}, namely if different alleles are favored in different geographic locations. Such selection, if strong enough, could lead to large variances in allele frequencies even if Nm were large, hence giving the incorrect impression that Nm is small.

Slatkin and Barton (1989) considered the problem of how the geometry of the population affects the estimate of Nm obtained using eq. (1.1). We showed that in a stepping-stone model of population structure, which represents the extreme in confined dispersal, the estimate of Nm, where m is the rate of migration to neighboring population, is of the correct order of magnitude but will depend somewhat on the distances separating populations sampled. We also considered a model in which individuals were continuously distributed in space (the "lattice model"), and showed that the estimates of Nm obtained were similar to Wright's neighborhood size. Thus, in a species in which individuals are grouped into discrete populations or in which they are continuously distributed, F_{ST} provides an estimate of the balance achieved between genetic drift and dispersal.

More recently, I have found that F_{ST} can provide more detailed information about isolation by distance (Slatkin 1991, 1993). The key is to compute F_{ST} for each pair of populations sampled and then use the resulting value to obtain an estimate of Nm from eq. (1.1). If that estimate is denoted by $\hat{M}$, then we obtain a value of $\hat{M}$ for each pair of locations in a sample. I have shown how $\hat{M}$ depends on geographic distance between pairs of locations. In a one-dimensional stepping-stone model, $\hat{M} \approx 4Nm/k$, where k is the distance separating two populations, and in a two-dimensional stepping-stone model, $\hat{M} \approx 4Nm/\sqrt{k}$ for relatively small distances. This relationship does not depend on the mutation rate at the locus of interest as long as it is small enough. Thus it is possible to achieve Wright's goal of predicting some features of the genetic structure from the demographic structure, namely, pairwise values of F_{ST} or equivalently $\hat{M}$ can be predicted from N and m even though other features, particularly the variance in frequency among locations and the correlation in allele frequencies, cannot be.

Takahata (1983) and Crow and Aoki (1984) showed that F_{ST} in an island model approaches its equilibrium value on a time scale determined by the smaller of N and $1/m$, which is consistent with Wright's result for the variance in allele frequency. Therefore, if m is relatively large, say 0.01 or larger, than F_{ST} will approach its equilibrium value in one hundred generations or so. That may still be longer than we would like, but it is better than other properties of the population such as the heterozygosity, which approaches its equilibrium value much more slowly (Nagylaki 1977).

More recently, I have considered the rate of approach to the equilibrium of F_{ST} in stepping-stone models (Slatkin 1993). I found that even though the overall time it takes for F_{ST} between pairs of populations to reach their equilibrium values is quite long, the pattern of F_{ST} values before the equilibrium is reached is also revealing. I considered a model in which there is a collection of empty sites that are colonized all at once at time τ in the past by populations descended from a single ancestral population. This model represents the very rapid spread of a species into a previously unoccupied area. Between time τ and the present, I assumed that there was gene flow between adjacent populations. I found that the equilibrium pattern of $\hat{M}$ as a function of distance would be observed between populations that are sufficiently close together, and that there would be no relationship between $\hat{M}$ and distance for populations that are separated by greater distances. In other words, the pattern of isolation by distance spreads with increasing values of τ. The diameter of the area in which the equilibrium pattern is detectable is roughly proportional to $\sqrt{Nm\tau}$.

There have been several simulation studies to test whether F_{ST} does lead

TABLE 1.1
Estimates of $4Nm$ in a simulated data set

Actual $4Nm$	G_{ST}	θ
0.512	0.558	0.766
5.12	5.603	5.227
51.2	30.768	54.646

Notes: A coalescent model was used to simulate the frequencies of neutral alleles in an island model of population structure. In each case, there were one hundred demes each with 12,800 individuals. Mutations occurred at a rate $\mu = 10^{-6}$ per generation to alleles not previously found in the population. Twenty-five diploid individuals were sampled from each of ten demes. The simulation method is the same as described by Hudson (1990) and in chapter 2. The ancestry of the genes sampled is simulated by computing the probabilities of coalescence and migration events in each generation. The simulation proceeded until only a single ancestor remained. Then mutation events were assumed to occur at rate μ per generation on each branch independently. The parameter values were chosen to match those of Slatkin and Barton (1989, table 1). The values of Nm and $N\mu$ are the same but the size of each deme had to be increased by a factor of 100 in order to allow the use of the coalescent approach. G_{ST} indicates that $4Nm$ was estimated from F_{ST} computed from Nei's (1973) G_{ST} statistic, and θ indicates that $4Nm$ was estimated from F_{ST} computed from Weir and Cockerham's (1984) $\hat{\theta}$ statistic. Estimates are based on the averages of ten independent sets of ten replicates, with each replicate representing a polymorphic locus.

to accurate estimates of Nm under conditions where it would be expected to. Slatkin and Barton (1989) presented some simulation results that showed that F_{ST} would lead to accurate estimates of Nm for moderate values of Nm ($Nm = 1$), but much less well for large Nm ($Nm = 10$) and for small values ($Nm = 0.1$). Recently, Cockerham and Weir (1993) have shown that F_{ST} provides much more accurate estimates of Nm for all values of Nm than claimed by Slatkin and Barton. Cockerham and Weir developed the appropriate sampling theory and demonstrated that Slatkin and Barton's results were not consistent with that theory. Cockerham and Weir emphasize that simulation results can be misleading if they are not checked against the appropriate analytic theory, a point with which I obviously have to agree. To illustrate the accuracy of F_{ST} in an island model, I carried out some new simulations, this time using a program based on a coalescent approach. The advantage of this program is that it is relatively easy to predict the average coalescence times of alleles drawn from the same and from different populations (Slatkin 1991). I verified that the mean coalescence times are in agreement with the theory. Table 1.1 shows some of the simulation results I obtained. They are in agree-

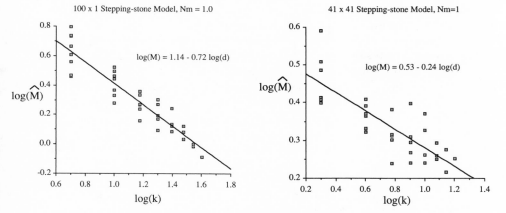

FIG. 1.1. Simulation results for one- and two-dimensional stepping-stone models. The results are based on single replicates of a stepping-stone model with 10,000 individuals in each deme. The infinite alleles model with $\mu = 10^{-6}$ was assumed. The simulation program used the coalescent approach, as described in table 1.1 and in chapter 2. In the one-dimensional stepping-stone model (*left*), nine demes with a spacing of four demes separating the nearest deme were sampled (i.e., demes 30, 35, 40, . . . , 70). In the two-dimensional model (*right*), the demes sampled were symmetrically placed on a line with one deme separating them [i.e., demes (12, 20), (14,20), . . . , (28,20)]. Estimates of $\hat{M}$ were obtained from eq. (1.1) using Weir and Cockerham's (1984) $\hat{\theta}$.

ment with those of Cockerham and Weir (1993) for comparable parameter values.

I have also used this simulation program to test the theory of isolation by distance described above. Figure 1.1 shows some typical results, which are described in more detail elsewhere (Slatkin 1993). The graphs show values of $\log(\hat{M})$ plotted as a function of $\log(k)$ where k is the distance separating two locations. The lines are the regression lines through the points. That is, values of a and b were chosen to provide the best fit of the equation $\log(\hat{M}) = a + b \log(k)$ to the simulation results. The predicted values b are −1 for the one-dimensional model and −1/2 for the two-dimensional model. The regression coefficients are somewhat smaller than expected because of the variation in $\hat{M}$ values about their expectation and the relatively small number of loci (ten) assumed in the model. Slatkin (1993) presents more extensive simulation results of this type and also some simulation results for the nonequilibrium model described above.

There are other indirect estimators of Nm, including the average frequency of private alleles and maximum likelihood methods. Slatkin and Barton (1989) discuss these methods in more detail. Maximum likelihood methods appear to be biased unless a rather large number of locations are included in the sample. The problems with the simulations in the Slatkin and Barton paper do not affect that conclusion. Two different maximum likelihood methods produced estimates of Nm that were too large by a factor of two for intermediate levels of gene flow ($Nm = 1$). Slatkin and Barton also examined a different estimator, based on the frequencies of "private"alleles (Slatkin 1985b). Although private alleles are in theory as reliable as F_{ST}, that method is likely to be more sensitive to errors in data collection and hence is probably less accurate in practice.

Examples

Numerous studies have applied both direct and indirect methods to natural populations. Using either method alone provides information about the species being studied, but in the absence of other knowledge about the species it is difficult to know what is actually learned. As I have emphasized, both classes of methods rely on assumptions that cannot be independently tested and each method yields different information. Studies in which both direct and indirect methods have been used provide some insight into population structure and illustrate the relationship between these two classes of methods. These studies suggest that species fall roughly into three categories. The first contains species that direct studies indicate have a high dispersal ability and indirect studies indicate that there are high levels of gene flow. Several bird species are of this type (Barrowclough and Johnson 1988). In most cases detailed estimates of dispersal distances are not available, but behavioral studies indicate that long-distance dispersal is common. Similarly, several species of marine invertebrates, including the common mussel, *Mytilus edulis*, have larvae that float in the plankton for a few months. Individual dispersal distances have never been measured but it appears that long-distance dispersal occurs frequently. Koehn, Milkman, and Mitton (1976) found little differentiation at most allozyme loci.

    The second category contains species for which both direct and indirect species indicate that there is very little dispersal. Several salamander species are of this type (Slatkin 1985b). One species, *Batracoseps campi*, is quite sedentary and has not been found to move even between adjacent populations (Yanef and Wake 1981). Electrophoretic studies indicate that there is essentially no gene flow between different populations of this species. A similar situation is found in prairie herb *Liatrix cylindricea* studied by Schaal (1975). She found substantial local differentiation of plants in adjacent quadrats within a study area of 18 m × 33 m. Although Schaal did not carry out a direct study, she argued that the dispersal of both pollen and seeds was restricted. Another example is the pocket gopher, *Thomomys bottae*, studied by Patton and Smith (1990). That species is very restricted in dispersal and has relatively high F_{ST} values when populations in different geographic areas are compared. The actual pattern of local differentiation in *T. bottae* is more complex, and it appears that in some areas a genetic equilibrium has not been reached (Slatkin 1993). In still another example, Roeloffs and Reichert (1988) found evidence of large genetic differences between colonies of the social spider *Agelena consociata* that is consistent with their observations of little or no dispersal among colonies in this species.

The third category, which in many ways is the most interesting, contains species in which direct studies indicate that dispersal occurs over very short distances, yet indirect studies indicate that substantial gene flow occurs over much longer distances. *Drosophila pseudoobscura* has been found to disperse a few kilometers even in very harsh environments (Coyne et al. 1982), yet the average Nm for populations from throughout the western United States is greater than 1. Furthermore, there is no evidence of isolation by distance, which would be expected given the limited dispersal (Lewontin 1974). Other species exhibit a similar pattern. The checkerspot butterfly *Euphydrias editha* disperses very little, yet indirect studies show evidence of gene flow over long distances (Slatkin 1985a). Caccone (1985) found that dispersal ability did not correlate well with apparent levels of gene flow in several cave-living species. Some species with restricted dispersal had moderate to high levels of gene flow.

There are two obvious explanations for the difference between direct and indirect estimates of gene flow. One is that dispersal is highly variable in time. Direct studies may miss rare dispersal events that are responsible for the high levels of gene flow found in indirect studies. That is probably true for some species because long-distance dispersal can be triggered by unusual ecological conditions. It is well known that animals that commonly do not move can move long distances when they are forced to. For other species, however, that explanation seems unlikely. Ehrlich and his coworkers (Ehrlich et al. 1975) have amassed considerable evidence that checkerspot butterflies from one geographic area could not interbreed with those in another area even if they could disperse that far. In that species, the breeding season in each locality is relatively short and tied to the phenology of the host plant. In different areas, different host plants are used, thus making the breeding seasons of different butterfly populations nonoverlapping. In *Drosophila pseudoobscura*, flies can travel much farther than a few kilometers, but it is difficult to imagine the conditions that would lead to enough dispersal to eliminate a pattern of isolation by distances over thousands of kilometers.

Another explanation is that these species are not yet at a genetic equilibrium under their current demographic conditions. That would be the case if, for example, the species had recently expanded its geographic range. We know this process can occur very quickly. *Drosophila subobscura* expanded its range by 1000 km in about ten years in both North and South America (Beckenbach and Prevosti 1986), although we cannot be certain that this species was not transported partly by humans. Numerous studies of invasions by other exotic species show the capacity for rapid range expansion (Elton 1958). The difficulty is that, unless there are good historical or fossil records, it is impossible to be sure of what the correct explanation for this pattern is. Nevertheless, a difference between direct and indirect estimates of gene flow does indicate that the species being studied is not at equilibrium, at least under the force of dispersal that can be measured with direct methods.

There is another possible category of species: those which direct methods indicate have high levels of gene flow and indirect methods indicate very restricted dispersal. I know of no examples of that type. This pattern could be caused by strong selection in favor of different alleles in different geographic locations at several of the genetic loci surveyed. There are of course abundant examples of clines caused by selection (Endler 1977), but that has not been found at more than one locus in most species. There is evidence in *Drosophila melanogaster* of parallel clines at four loci (Oakeshott et al. 1981, 1982), but in that species over a hundred loci have been surveyed. Most loci provide evidence of higher levels of gene flow (Singh and Rohmberg 1987), and that species, like *D. pseudoobscura*, are probably not at a genetic equilibrium. The absence of species in this fourth category lends support to the idea that spatially varying selection is not greatly distorting indirect estimates of gene flow for most loci.

Conclusions

There are two main conclusions from my analysis of the geographic structure of species. The first is that indirect methods for estimating levels and patterns of gene flow using Wright's F_{ST} statistic do provide reliable estimates of the average value of Nm and can indicate the extent of isolation by distance. F_{ST} is relatively insensitive to mutation rates and selection coefficients, making it possible to combine information from different loci. The second conclusion is that when both direct and indirect methods are applied to the same species, it is sometimes possible to determine whether a species is at a genetic equilibrium under current levels of gene flow. There appears to be no generalization about the role of gene flow. Some species are at an equilibrium under high levels of gene flow and others are at an equilibrium under very low levels of gene flow. Other species appear to be not at an equilibrium either because of occasional episodes of very high levels of gene flow or because of a recent range expansion.

Whether or not a species is at a genetic equilibrium is of interest in its own right. That is one more bit of information and one that cannot be obtained in other ways. It is also of interest for understanding the evolutionary potential of such species. Gene flow can constrain evolution by preventing the adaptation to local conditions. If a species is not at an equilibrium under current levels of gene flow, then selection is unconstrained by gene flow and local adaptations may evolve more readily.

References

Barrowclough, G. F., and N. K. Johnson. 1988. Genetic structure of North American birds. In *Acta XIX Congressus Internationalis Ornithologici*, ed. H. Ouellet, vol. 2, pp. 1630–1638. University of Ottawa Press, Ottawa.

Bateman, A. J. 1950. Is gene dispersion normal? *Heredity* 4: 353–363.

Beckenbach, A., and A. Prevosti. 1986. Colonization of North America by the European species *Drosophila subobscura* and *D. ambigua*. *Amer. Midland Nat.* 115: 10-18.

Caccone, A. 1985. Gene flow in cave arthropods: A qualitative and quantitative approach. *Evolution* 39: 1223–1235.

Cockerham, C. C., and B. S. Weir. 1993. Estimation of gene flow from F-statistics. *Evolution*, in press.

Coyne, J. A., I. A. Boussy, T. Prout, S. H. Bryant, J. S. Jones, and J. A. Moore 1982. Long distance migration of Drosophila. *Amer. Nat.* 119: 589–595.

Crow, J. F. and K. Aoki. 1984. Group selection for a polygenic behavioral trait: Estimating the degree of population subdivision. *Proc. Natl. Acad. Sci. USA* 81: 6073–6077.

Ehrlich, P. R., R. R. White, M. C. Singer, S. W. McKechnie, and L. E. Gilbert. 1975. Checkerspot butterflies: A historical perspective. *Science* 188: 221–228.

Ellstrand, N. C., and D. L. Marshall. 1985. Interpopulation gene flow by pollen in the wild radish, *Raphanus sativus*. *Amer. Nat.* 126: 606–616.

Elton, C. S. 1958. *The Ecology of Invasions by Animals and Plants*. Methuen, London.

Endler, J. A. 1977. *Geographic Variation, Speciation, and Clines*. Princeton University Press, Princeton, N.J.

Ewens, W. J. 1989. The effective population sizes in the presence of catastrophes. In *Mathematical Evolutionary Theory*, ed. M. W. Feldman, pp. 9–25. Princeton University Press, Princeton, N.J.

Haldane, J.B.S. 1930. A mathematical theory of natural and artificial selection. Part 4, Isolation. *Proc. Cambridge Philos. Soc.* 26: 220–230.

Handel, S. N. 1982. Dynamics of gene flow in an experimental population of *Cucumis melo* (Cucurbitaceae) *Amer. J. Bot.* 69: 1538–1546.

Hudson, R. R. 1990. Gene genealogies and the coalescent process. *Oxford Surv. Evol. Biol.* 7: 1–44.

Kimura, M., and J. F. Crow. 1963. The measurement of effective population number. *Evolution* 17: 279–288.

Kimura, M., and G. H. Weiss. 1964. The stepping stone model of population structure and the decrease of genetic correlation with distance. *Genetics* 49: 561–576.

Koehn, R. K., R. Milkman, and J. B. Mitton. 1976. Population genetics of marine pelecypods. IV. Selection, migration and genetic differentiation in the blue mussel, *Mytilus edulis*. *Evolution* 30: 2–32.

Lewontin, R. C. 1974. *Genetic Basis of Evolutionary Change*. Columbia University Press, New York.

Malécot, G. 1968. *The Mathematics of Heredity*. Freeman, San Francisco.

Nagylaki, T. 1975. Conditions for the existence of clines. *Genetics* 80: 595–615.

Nagylaki, T. 1977. Decay of genetic variability in geographically structured populations. *Proc. Natl. Acad. Sci. USA* 74: 2523–2525.

Nei, M. 1973. Analysis of gene diversity in subdivided populations. *Proc. Natl. Acad. Sci. USA* 70: 3321–3323.

Oakeshott, J. G., G. K. Chambers, J. B. Gibson, and D. A. Willcocks. 1981. Latitudinal relationships of esterase-6 and phosphoglucomutase gene frequencies in *Drosophila melanogaster*. *Heredity* 47: 385–296.

Oakeshott, J. G., J. B. Gibson, P. R. Anderson, W. R. Knibb, D. G. Anderson, and G. K. Chambers. 1982. Alcohol dehydrogenase and glycerol-3-phosphate dehydrogenase clines in *Drosophila melanogaster* on different continents. *Evolution* 36: 86–96.

Patton, J. L., and M. F. Smith. 1990. The evolutionary dynamics of the pocket gopher *Thomomys bottae*, with emphasis on California populations. *University of California Publications in Zoology*, vol. 123. University of California Press, Berkeley.

Roeloffs, R., and S. E. Reichert. 1988. Dispersal and population-genetic structure of the cooperative spider, *Agelena consociata*, in a west African rainforest. *Evolution* 42: 173–183.

Schaal, B. A. 1975. Population structure and local differentiation in *Liatris cylindracea*. *Amer. Nat.* 109: 511–528.

Singh, R. S., and L. R. Rohmberg. 1987. A comprehensive study of genic variation in natural populations of *Drosophila melanogaster*. *Genetics* 115: 313–322.

Slatkin, M. 1985a. Gene flow in natural populations. *Ann. Rev. Ecol. Syst.* 16: 393–430.

Slatkin, M. 1985b. Rare alleles as indicators of gene flow. *Evolution* 39: 53–65.

Slatkin, M. 1991. Inbreeding coefficients and coalescence times. *Genet. Res.* 58: 167–175.

Slatkin, M. 1993. Isolation by distance in equilibrium and nonequilibrium populations. *Evolution* 47: 264–279.

Slatkin, M., and N. H. Barton. 1989. A comparison of three indirect methods for estimating average levels of gene flow. *Evolution* 43: 1349–1368.

Takahata, N. 1983. Gene identity and genetic differentiation of populations in the finite island model. *Genetics* 104: 497–512.

Weir, B. S., and C. C. Cockerham. 1984. Estimating F-statistics for the analysis of population structure. *Evolution* 38: 1358–1370.

Wright, S. 1931. Evolution in Mendelian populations. *Genetics* 16: 97–159.

Wright, S. 1943. Isolation by distance. *Genetics* 28: 114–138.

Wright, S. 1951. The genetical structure of populations. *Ann. Eugenics* 15: 323–354.

Wright, S. 1969. *Evolution and Genetics of Populations*, vol. 2, *The Theory of Gene Frequencies*. University of Chicago Press, Chicago.

Yanef, K. P., and D. B. Wake. 1981. Genic differentiation in a relict desert salamander, *Batrachoseps campi*. *Herpetologica* 37: 16–28.

2

Cladistic Analysis of DNA Sequence Data from Subdivided Populations

NEW BIOCHEMICAL methods for examining genes provide new kinds of data for analysis. Traditional population genetics has been formulated in terms of allele frequencies, and data obtained using gel electrophoresis provided vast amounts of data to be analyzed with traditional methods. In chapter 1 I discussed the use of one classical method to estimate average levels of gene flow. Restriction enzymes and DNA sequencing have provided much more detailed information about both coding genes and noncoding segments of DNA. Originally these new methods were used to provide comparisons of genes in different species, but more recently they have provided information about variation at the molecular level within species. With information about DNA sequences, it is possible to say not only whether two genes are the same or different but also how different they are. The challenge for population geneticists has been to find ways to make use of this additional information. In this chapter, I will describe the efforts of my collaborators and me to use DNA sequence data to learn about the magnitude and pattern of gene flow.

Sequence Data

The kind of data I will be concerned with here is a list of sequences of the same region of the genome in different individuals of the same species. For technical reasons, the mitochondrial genome in animals has been the most intensively studied, but within-species surveys of nuclear genes in both plants and animals are also being conducted. The polymerase chain reaction (PCR) method for amplifying small amounts of DNA for sequencing promises to vastly increase the amount of such data in the near future. Once primers are obtained for a particular region of the genome, it is relatively easy to sequence genes from several individuals of the same species.

During the 1980s, most of the information about within-species variation in DNA sequences came from locating polymorphic restriction sites for several restriction enzymes, which provides partial information about the sequence. In general, 6-base recognizing restriction enzymes (6-cutters) provide much less detailed information than do 4-base recognizing enzymes (4-cutters). In one of the most detailed of such studies, Kreitman and Aguadé (1986) estimated that their 4-cutters could detect all insertions and deletions and approximately 20% of the differences in base pairs in the ADH region of *Drosophila melanogaster*.

Gene Trees

To understand how data obtained from polymorphic restriction sites and complete sequences can be used, we need to consider what such data can reveal about the organisms being studied. Consider a single genetic locus, that is, a short segment of DNA that can be identified in each individual sampled from a population. For the purposes of this discussion, it does not matter whether this DNA is coding or noncoding. A convenient way to think about data is to imagine the history of the sample of interest. Because we are necessarily concerned with finite populations, homologous genes of the kind we are discussing are ultimately descended from a common ancestral gene. I will ignore gene conversion, though the same ideas can be extended to cover it. The history of genes in the sample can be represented by a cladogram with each node indicating that two (or more) genes are descended from a single ancestor in that generation. To be consistent with other usage in population genetics, I will call the cladogram representing the history of a sample of genes a "gene tree," and I will say that each node represents a "coalescence event" or simply a "coalescence." An example of a gene tree is shown in figure 2.1. A gene tree contains two kinds of information: the topological structure of the tree that indicates which genes coalesced with which other genes, and the branch lengths that indicate times between coalescence events.

The gene tree itself simply represents the pattern of ancestry of a set of genes. It tells us nothing about their DNA sequences or their phenotypic effects. What it does tell us is that mutations that occur on a branch will propagate by inheritance to all descendant genes. Therefore, differences between genes must be attributable to mutations that occurred in different lineages after their common ancestor was present. Similarities between genes are attributable either to common descent or to the same mutation occurring independently in different lineages after their common ancestor was present.

Although I have been discussing gene trees as if a single gene tree describes each locus, that is not true if there is any recombination. Figure 2.2 illustrates the effect of recombination on the ancestry of two halves of a locus. If we imagine tracing the ancestry of a locus backward

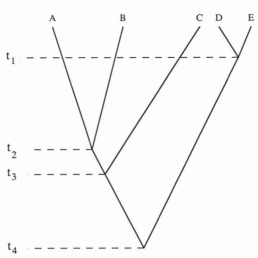

FIG. 2.1. An example of a gene tree representing the history of five genes (A–E) sampled from a population. The t_i indicate the times in the past that different coalescence events occurred.

in time, both halves, A and B, are found in the same individual in each generation. If a recombination event occurs, that means that A and B were on different homologous chromosomes before meiosis in that generation, and then on the same chromosome that was subsequently passed on to an offspring. Therefore, the chromosomes carrying A and B have different ancestries before the recombination event and hence are described by different gene trees. These two trees are not independent, however.

Consider the lineages of A and B before the recombination event. Both lineages will have to coalesce with other A and B regions in the sample. If there is only one recombination event in the history of the sample, then as soon as the A region in the recombinant coalesces with another lineage ancestral to the sample, it will follow the gene tree of that sample. Similarly, once the B region of the recombinant coalesces with another lineage it will also follow the gene tree of that sample. Before the recombination event, A and B are on different chromosomes and so do not have to coalesce with the same lineage. Consequently, the effect of a single recombination event is to make the gene trees of the A and B region differ by at most one "branch swap" (Dubose et al. 1988; Hein 1990). In other words, the gene tree of A can be obtained from the gene tree of B by cutting one branch and attaching it elsewhere. Each recombination event can be represented by one branch swap.

Mitochondrial DNA (mtDNA) in animals and chloroplast DNA (cpDNA) in plants apparently undergo no recombination, so a single gene tree can represent the entire genome. For nuclear genes that can

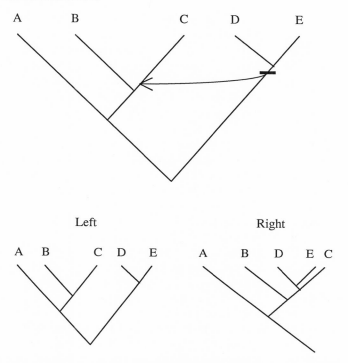

Fig. 2.2. An illustration of the effect of a single recombination event in the history of five genes in a sample. The top diagram indicates that a recombination event occurred in the gene that was ancestral to D and E. The right half of that gene then coalesced with the ancestor to C before the coalescence event with B. The two diagrams at the bottom indicate the gene trees of the left and right halves of the gene.

undergo recombination, there is no single gene tree. Instead, in a sample of genes from different individuals, some segments of DNA have not undergone recombination and are likely to have had only one recombination event between segments on either side. Instead of a single gene tree, there is a network of gene trees, each differing by a branch swap from its neighboring gene tree (Hein 1990).

Inferring Gene Trees

Because the history of a nonrecombined segment of DNA determines which mutations are shared by common ancestry, it is possible to use information about the sequence to infer the gene tree. Felsenstein (1988) reviews different ways to use sequence data to infer gene trees. Different methods of inference, including maximum parsimony and various distance methods, rely implicitly or explicitly on assumptions about the process of mutation. I will not be concerned with the method of inference but assume that by some means it is possible to infer the gene tree of the segment of the genome that is of interest. Given the difficulties inherent in inferring trees, it is probably best to use two or more methods to see if different trees are obtained and to see what conclusions follow from using different trees. Some methods for inferring trees provide only the topology and others, especially distance methods, provide both the topology and the branch lengths.

As a practical matter, any method for inferring gene trees requires enough variation to detect mutations that have occurred on each of the branches. If there is insufficient variation, it will not be possible to resolve all the nodes completely. The more information there is about the sequence, the more likely it is that there will be sufficient variation. The complete sequence provides the most information, but if the segment of nonrecombining DNA is long enough, then polymorphic restriction sites can provide enough information at least to partially resolve the gene tree. Avise et al. (1987) describe several studies of mtDNA using restriction enzymes.

From our current knowledge of recombination rates in nuclear genes in *Drosophila melanogaster* at least, it seems impossible to infer the gene tree for each nonrecombining segment of DNA. Estimates of the recombination rates between adjacent base pairs in the *Rosy* locus that codes for xanthine dehydrogenase are approximately 2×10^{-8} per generation, while estimates of the the mutation rate per base pair are approximately 10^{-9} per generation (Hudson and Kaplan 1988). This means that, in this region of the chromosome in this species, there are likely to be several recombination events separating polymorphic sites. The situation is different in species in which there is little recombination. Milkman and Stoltzfus (1988) could distinguish gene trees of different parts of the *E. coli* genome.

Using Gene Trees to Estimate Levels of Gene Flow

John Avise and his coworkers have been instrumental in promoting the idea that the gene tree can provide insight to population structure (Avise et al. 1987; Avise 1989). Motivated by Avise's discussions of this topic, Maddison and I have developed a method that uses part of the information in a gene tree to infer the average level of gene flow (Slatkin and Maddison 1989). Our method was developed to analyze mtDNA data, so we were not initially concerned with recombination. Our approach is to assume that the gene tree is known and then to treat the geographic location from which each gene was sampled as a character that evolved on the

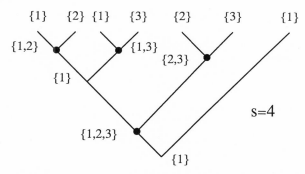

FIG. 2.3. An illustration of how to compute s, the minimum number of migration events in the history of seven genes sampled from three geographic locations, numbered 1, 2, and 3. The braces indicate the set of possible locations of each gene. The dark circles indicate that a migration event must be assumed at that node. The rule is to assign the state set of the ancestor to be the intersection of two sets if that intersection is not empty. If the intersection is empty, then assign the state set of the ancestor to be the union of the two sets and count one additional migration event.

tree. A change in the state of that character represents a migration event. Given the tree and the geographic locations, it is possible to compute the minimum number of migration events that must have occurred.

The procedure is simple and is illustrated in figure 2.3 for samples from three populations. The state of a branch is represented by a set. At the tips, each set contains only a single element, the geographic location from which that gene was sampled. Internal nodes representing ancestral genes may have ambiguous states. Working down the tree from the tips, the state of an internal node is the intersection (in the set theoretic sense) of the branches that coalesce if that intersection is not empty. If it is empty, then one migration event is counted and the state of the node is the union of the states of the two descendants. Fitch (1971) proved that this algorithm yields the minimum number of state changes.

We denoted the minimum number of migration events by s. It is reasonable to suppose that larger values of s indicate more gene flow. We can see that more clearly by considering the smallest possible value of s. If n geographic locations are sampled, then s must be at least $n - 1$. If $s = n - 1$, it means that genes in each location have the same common ancestor, so there is complete concordance between the gene tree and the locations from which genes were sampled, as illustrated in figure 2.4. As Avise et al. (1987) discussed, that pattern is what would be expected if there were little or no gene flow. I considered this problem and found a way to put an upper bound on the amount of gene flow that is consistent with this pattern (Slatkin 1989). A value of s greater than $n - 1$ is consistent with higher levels of gene flow. The question is how much higher.

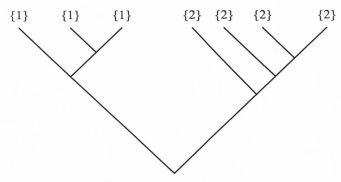

{1}   {1}   {1}   {2}   {2}   {2}   {2}

FIG. 2.4. An example of a gene tree of seven genes sampled from two geographic locations in which there is complete concordance between the gene tree and the geographic locations. In this case, s, the minimum number of migrations events, is one.

Island Model of Population Structure

To find the relationship between the amount of gene flow and the s value, Maddison and I wrote a program that simulates the history of a sample of genes from a subdivided population. Our model assumed a population of haploid organisms in an island model of population structure. We assumed that there were d islands with an immigration rate of m for every island. Each immigrant had a probability of $1/(d-1)$ of coming from each other's island. We assumed each island contained N randomly mating, monoecious individuals.

In our simulations, we used a coalescent approach, described in detail by Hudson (1990), to produce a sample history of a sample of genes. We started with n individuals sampled from each of two demes in the island model and simulated the processes of migration and mating in each generation backward in time until only a single ancestor of the $2n$ genes remained. A convenient feature of this approach to simulating genetic evolution is that it naturally leads to a gene tree of the sample. In each generation in the past, the state of the population is described by the number of ancestral genes in each of the d demes, $i_1, \ldots, i_d$. The probability that there is a coalescence event in deme j is approximately $i_j(i_j - 1)/(2N)$ in each generation, and the probability that one of the i_j genes is an immigrant from some other population is mi_j. We assumed that N was sufficiently large and m sufficiently small that there would be at most one migration or coalescence event in each generation.

For given parameter values, each replicate simulation produced a single gene tree, which represented a possible pattern of ancestry under our assumptions. For this gene tree we computed s, the minimum number of

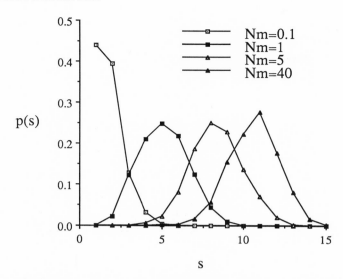

FIG. 2.5. The distribution of s, $p(s)$ in 1,000 replicate simulations of an island model. In each case, a model with ten demes with 10,000 individuals per deme was assumed, and sixteen genes from each of two demes were sampled. (From Slatkin and Maddison 1989)

migration events. By running a large number of replicates, we determined the approximate distribution of s for each set of parameter values. It was convenient that this distribution depended only on the product Nm and not on either parameter separately. Furthermore, it was almost independent of d for $d \geq 10$. Therefore, we can understand the pattern of the results by choosing $d = 10$ and varying Nm. Some examples are shown in figure 2.5.

The distribution of s and the value of $\bar{s}$, the average of s, depend also on the number of individuals sampled from each geographic locations. We found that $\bar{s}$ was most sensitive to the smaller sample sizes (Slatkin and Maddison 1989). Little extra information was gained by having very large samples from one location and small samples from another. If all the sample sizes are increased by a factor c, then $\bar{s}$ is increased by that factor as well.

Figure 2.5 confirms our intuition about the dependence of s on Nm. If $Nm \ll 1$, $s = 1$ most of the time, while if Nm is 1 or larger, larger values of s are found. The variance of s is approximately constant if $Nm \geq 1$, so the average of s, $\bar{s}$, provides a convenient summary of the dependence of this distribution on s. Figure 2.6 shows $\bar{s}$ as a function of Nm.

The results in figure 2.6 suggest that s obtained for a particular data set

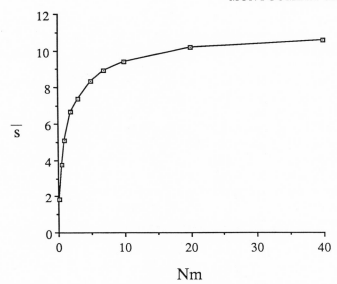

FIG. 2.6. The mean of s, $\bar{s}$, plotted as a function of Nm in an island model. The parameter values are the same as in figure 2.5. (From Slatkin and Maddison 1989)

could be used to estimate Nm. For equal sample sizes from each of two locations, figure 2.6 could be used, but in general more accurate estimates can be obtained using a simulation program that Maddison and I wrote and will distribute on request. This program also provides confidence limits on the estimate of Nm. For example, if sixteen individuals were sampled from each of two locations and s were found to be 5, our program estimated Nm to be 2.0 with 95% confidence limits of 0.3 and 8.0.

These results show that s is a potentially useful measure of the average level of gene flow if a gene tree can be inferred from the data. Two obvious questions arise about its use: What is the effect of geographic structure, and what is the effect of recombination? I will discuss each of these in turn.

Isolation by Distance

The island model of population structure represents the extreme in long-distance migration; each local population is equally accessible to the other. In contrast, the stepping-stone model, in which gene flow occurs between only neighboring populations, represents the extreme in confined gene flow. As I discussed in the previous chapter, restricted dispersal leads to isolation by distance. Maddison and I considered this problem and showed how our measure of effective migration rate in the island model could be used to detect isolation by distance (Slatkin and Maddison 1990).

For samples from any two populations, the gene tree for those samples will lead to a value of s, the minimum number of migration events between those two populations. The calculation of s assumes nothing about the actual dispersal pattern or the geographic relationship of those two populations. We can then use the method described in the previous section and obtain an estimate of Nm for this value of s. The resulting value of Nm is the amount of gene flow in an island model that would have to be present in order to explain that value of s. We defined the value of Nm obtained this way to be $\hat{M}$, which depends on s and the sample sizes and which is a measure of the net gene flow between the two populations sampled. The value of $\hat{M}$ indicates the degree of genetic similarity of these two populations. I am using the same notation as in chapter 1 because the logic is exactly the same. In both cases, $\hat{M}$ indicates the value of Nm in an island model that would account for the observed degree of genetic similarity. The only difference is the measure of genetic similarity. In the previous chapter, in which allozyme data were discussed, similarity was defined in terms of F_{ST}, while in this chapter genetic similarity is defined in terms of s.

Maddison and I found that the dependence of $\hat{M}$ on the geographic distance separating pairs of populations was very similar to what I found for $\hat{M}$ based on allozyme data. Figures 2.7 shows the results for a one- and two-dimensional stepping-stone model. As in the previous chapter, $\log(\hat{M})$ is approximately a linear function of $\log(k)$, where k is the distance separating populations sampled. In a one-dimensional model, the slope is approximately -1, and it is approximately $-1/2$ in a two-dimensional model (Slatkin and Maddison 1990).

We also considered a model of a continuously distributed population in which each individual was located at the points on a two-dimensional lattice. We assumed that the parents of an individual were chosen from different distributions with specified variances, σ^2. We then assumed that individuals were sampled randomly from small quadrats whose centers were separated by a specified distance. We treated samples from each quadrat as if they were from a single population so we could compute a value of s between each pair of quadrats. The results were very similar to those for the stepping-stone model. If the lattice was long and thin, resembling a one-dimensional stepping-stone model, $\log(\hat{M})$ decreased roughly with $-\log(k)$, while if the lattice was square, $\log(\hat{M})$ decreased roughly with $-(1/2)\log(k)$. Some results are shown in figure 2.8.

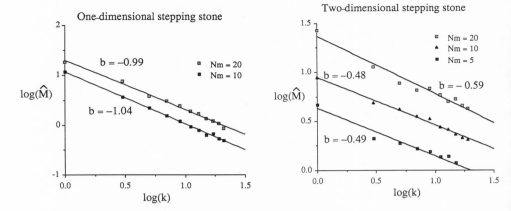

FIG. 2.7. The dependence of the $\log_{10}(\hat{M})$ on $\log_{10}(k)$, where k is the linear distance separating pairs of locations sampled in a stepping-stone model. At *left*, a 100×1 array of demes was assumed, and at *right*, a 21×21 array was assumed. In both cases, demes were assumed to be of size 10,000, and sixteen genes sampled from each of two populations were simulated. The value of b is the regression coefficient of $\log(\hat{M})$ on $\log(k)$. (From Slatkin and Maddison 1990)

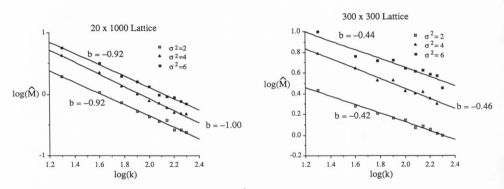

FIG. 2.8. The dependence of the $\log_{10}(\hat{M})$ on $\log_{10}(k)$, where k is the linear distance separating pairs of locations sampled in a lattice model. Individuals were located at each point in a two-dimensional lattice of specified dimensions. Dispersal distances for each parental gamete were assumed to be drawn from a circular exponential distribution with variance σ^2. In each replicate sixteen individuals were sampled without replacement from two 5×5 quadrats whose centers were separated by k lattice points. The value of b is the regression coefficient of $\log(\hat{M})$ on $\log(k)$. (From Slatkin and Maddison 1990)

Testing Hypotheses about Isolation by Distance

The relatively simple dependence of $\hat{M}$ on geographic distance suggested that our results could be used to test different hypotheses about isolation by distance. If several locations have been sampled, then a value of s and hence $\hat{M}$ could be obtained for each pair of locations. For a given set of locations, there could be several hypotheses about the actual pathways of gene flow. Each hypothesis about gene flow would be represented by a different matrix of geographic distances separating these populations. The example we discussed in detail was of nine populations arranged in a square, as shown in figure 2.9 (Slatkin and Maddison 1990). One hypothesis about these populations is that dispersal occurs in two dimensions, so that geographic distance between any two populations is just the Euclidean distance separating them. Another hypothesis is that dispersal is confined to the dark band indicated in the figure, which could represent the course of a river. In that case, the distance between populations is the distance measured along the river. The biological question is whether or not the river is the primary pathway of gene flow.

Given these two hypotheses, the procedure is to regress $\hat{M}$ against distance using each of these two distance matrices. The results for one case are shown in figure 2.10. In this case, we would conclude that the linear distance model fits much better than does the model that assumes dispersal is confined to the river. In one case the curve is nearly linear with a slope near –1/2, and in the other case the curve is not even monotonic and the slope of the regression line is too small to be consistent with a model of isolation by distance. Unfortunately, the results in this one case were somewhat fortuitous, and for these sample sizes there is not much power to distinguish the two hypotheses. More power is gained by taking samples in a line, thereby increasing the number of distance classes (Slatkin and Maddison 1990). There is considerable power, though, for rejecting the hypothesis that samples are from a species in which dispersal distances are larger than the longest distances between sampling locations. In that case, no apparent isolation by distance would be found.

Another hypothesis about gene flow that can be tested using this method is whether a species is at a genetic equilibrium. The approach is, not surprisingly, the same as what I described in chapter 1. When there is direct evidence of restricted dispersal, then the absence of any pattern of isolation by distance suggests that the species has not been in its present location long enough for the restriction of gene flow to affect the apparent migration rates. Maddison and I analyzed data on human mtDNA from

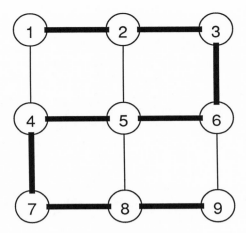

FIG. 2.9. An example of two possible hypotheses about gene flow among nine geo-graphic locations. The thinner lines indicate that gene flow could occur between near-est neighbors, which would mean that the appropriate measure of geographic distance is a linear distance. The thicker lines indicate that gene flow could be constrained to particular paths, which would mean that the appropriate measure of geographic dis-tance between pairs of locations is the distance measured along the thick lines.

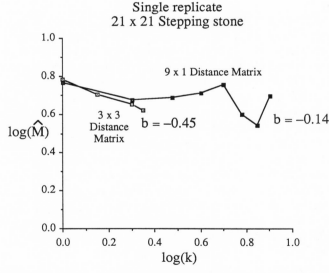

FIG. 2.10. An illustration of how the hypotheses represented in figure 2.9 might be tested. Sample data were generated by simulating the ancestry of sixteen genes sampled from each of nine locations. Then, values of s and $\hat{M}$ were computed for each pair of locations. The points show the results from plotting $\log(\hat{M})$ against geographic dis-tance for each of the two models of distance between pairs of locations. The 3×3 distance matrix assumes that dispersal occurs between nearest neighbors and the 9×1 distance matrix assumes that dispersal is confined to the thick line in figure 2.9. In each case, b is the regression coefficient of $\log(\hat{M})$ against $\log(k)$, where k is the distance. (From Slatkin and Maddison 1990)

five races and found that there was no pattern of isolation by distance (Slatkin and Maddison 1990). This result is in agreement with other evidence suggesting relatively recent dispersal of modern humans.

Recombination

The results described so far were derived under the assumption that no recombination in the region of DNA was sampled, as is appropriate for mtDNA in animals. As I have discussed, if there is recombination, then there is no single gene tree that describes the history of the segment of DNA that is of interest. It is possible to proceed, however, as if that were not true and use the same method to infer a single gene tree anyway. From that tree a value of s could still be found, and that value could then be used to estimate Nm. Intuition suggests that this approach would yield a reliable estimate when recombination rates are small enough but not when they are larger. Surprisingly that is not the case. Hudson, Slatkin, and Maddison (1992) carried out a simulation study that shows that the performance of the cladistic method is relatively insensitive to recombination.

We simulated an island model and sampled sixteen individuals from each of two populations. We then used both a maximum parsimony method and a clustering method (UPGMA) on each sample data set to infer a gene tree, from which we computed s. Figure 2.11 shows some of our results. The average value of s is not sensitive to the value of Nr, the product of the population size and the recombination rate. This means that even with recombination, s could be used to estimate Nm. We also applied a method suggested by Lynch and Crease (1990), which treats each site as an allele and computes a value of F_{ST} from what becomes in effect allele frequency data. That value of F_{ST} is then used to estimate Nm. When there is no recombination, our results show that F_{ST} leads to relatively poor estimates of Nm. As the recombination rate increases, however, their method improves while the cladistic method does not. The explanation is that recombination is making sites become more independent, so the F_{ST} value is in effect based on more independent loci.

Conclusion

I had two goals in this chapter. One was to show that quite different approaches are needed to analyze DNA sequence data than have been developed to analyze allele frequency data. That is particularly true when there is no recombination. In that case, the results discussed here support

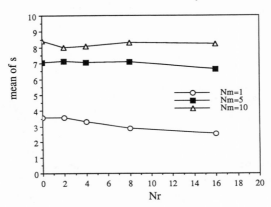

s Calculated from Parsimony trees

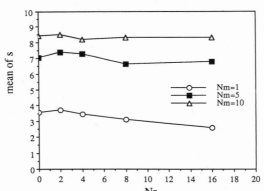

s Calculated from UPGMA trees

FIG. 2.11. Values of $\bar{s}$ computed from the inferred gene tree of sixteen genes sampled from an island model. Recombination was assumed to occur at a rate r per chromosomal segment. The simulations were of a segment of the infinite sites models, so no site could mutate more than once. The simulations were tailored so that there were 128 polymorphic sites per sample. An island model with ten demes and 10,000 individuals per deme was assumed. Each point represents the average of one hundred replicates. In the top graph, the gene tree was inferred using maximum parsimony, and in the bottom graph the gene tree was inferred from the same data using UPGMA. (Based on simulation results from Hudson, Maddison, and Slatkin 1992)

the ideas of Avise and his coworkers that a phylogenetic approach is necessary to take advantage of the information in the data. The second goal was to show that there is already one class of methods that can use phylogenetic information to infer levels and patterns of gene flow. This is not the only possible method or necessarily the best. The cladistic method that Maddison and I developed does not make any use of branch lengths, only the topology of the gene tree. Instead, I regard this work as a first attempt to analyze what promises to be a rich area of theoretical analysis, and one that will be increasingly important to people concerned with the geographic structure of natural populations.

References

Avise, J. 1989. Gene trees and organismal histories: A phylogenetic approach to population biology. *Evolution* 43: 1192–1208.

Avise, J. C., J. Arnold, R. M. Ball, E. Bermingham, T. Lamb, J. E. Neigel, C. A. Reeb, and N. C. Saunders. 1987. Intraspecific phylogeography: The mitochondrial DNA bridge between population genetics and systematics. *Ann. Rev. Ecol. Syst.* 18: 489–522.

Dubose, R. F., D. E. Dykhuizen, and D. L. Hartl. 1988. Genetic exchange among natural isolated bacteria: Recombination within the *pho*A locus of *Escherichia coli. Proc. Natl. Acad. Sci. USA* 85: 7036–7040.

Felsenstein, J. 1988. Phylogenies from molecular sequences: Inference and reliability. *Ann. Rev. Genet.* 22: 521–565.

Fitch, W. M. 1971. Toward defining the course of evolution: Minimum change for a specific tree topology. *Syst. Zool.* 20: 406–416.

Kreitman, M., and M. Aguadé. 1986. Genetic uniformity in two populations of *Drosophila melanogaster* as revealed by filter hybridization of 4-nucleotide-recognizing restriction enzyme digests. *Proc. Natl. Acad. Sci. USA* 83: 3562–3566.

Hein, J. 1990. Reconstructing evolution of sequences subject to recombination using parsimony. *Math. Biosci.* 98: 185–200.

Hudson, R. R. 1990. Gene genealogies and the coalescent process. In *Oxford Surveys in Evolutionary Biology*, ed. D. J. Futuyma and J. Antonovics, vol. 7, pp. 1–44. Oxford University Press, Oxford.

Hudson, R. R., and N. L. Kaplan. 1988. The coalescent process in models with selection and recombination. *Genetics* 120: 831–840.

Hudson, R. R., M. Slatkin, and W. P. Maddison. 1992. Estimation of levels of gene flow from DNA sequence data using cladistic and pairwise methods. *Genetics* 132: 583–589.

Lynch, M., and T. J. Crease. 1990. The analysis of population survey data on DNA sequence variation. *Mol. Biol. and Evol.* 7: 377–394.

Milkman, R., and A. Stoltzfus. 1988. Molecular evolution of the *Escherichia coli* chromosome. II. Clonal segments. *Genetics* 120: 359–366.

Slatkin, M. 1987. The average number of sites separating DNA sequences drawn from a subdivided population. *Theoret. Pop. Biol.* 32: 42–49.

Slatkin, M. 1989. Detecting small amounts of gene flow from phylogenies of alleles. *Genetics* 121: 609–612.

Slatkin, M., and W. P. Maddison. 1989. A cladistic measure of gene flow from the phylogenies of alleles. *Genetics* 123: 603–613.

Slatkin, M., and W. P. Maddison. 1990. Detecting isolation by distance using phylogenies of genes. *Genetics* 126: 249–260.

3

The Evolution of Phenotypic Plasticity:
What Do We Really Know?

Introduction

Individuals of the same species that develop in different environments may differ considerably in ecologically important phenotypic characters. This phenotypic differentiation between groups sampled from various habitats may be due either to genetic differences among the groups (as shown in chapter 4) or to purely environmental effects on the phenotypes of individuals. For a given genotype, a phenotypic change caused only by a change in the environment is called "phenotypic plasticity." The set of phenotypes produced by a genotype over environments is termed its "norm of reaction."

The study of phenotypic responses to the environment has a long history (classic papers include Woltereck 1909, Gause 1947, Schmalhausen 1949, and Bradshaw 1965). Recently there has been a resurgence of interest in phenotypic plasticity, with a spate of new theoretical and empirical studies (reviews in Schlichting 1986, Via 1987, and Stearns 1989). Although there is general agreement at the simplest descriptive level about what constitutes phenotypic plasticity (phenotypic change across environments), it has been harder to reach a consensus on how plasticity can best be measured in natural populations (review in Via 1987). Agreement on mechanisms by which phenotypic plasticity might evolve has proved even more difficult to achieve (Schlichting and Levin 1984, 1986; Schlichting 1986, 1989; Via 1987; Stearns 1989; Scheiner and Lyman 1989, 1991).

One idea that has played a large role in recent discussions is that plasticity and trait values are independent of one another, and that this independence suggests the existence of "genes for plasticity" that are separate from the loci that influence trait means (Bradshaw 1965; Schlichting and Levin 1986; Jinks and Pooni 1988; Scheiner and Lyman 1989, 1991). In this chapter, I will attempt to illustrate that, as it is presently articulated, this view of the genetic basis of phenotypic plasticity is not only unnecessary, it is at times inconsistent with current understanding of the genetics and evolution of the ecologically important characters that exhibit phenotypic plasticity.

The approach to the study of phenotypic plasticity that is discussed here originated in the formulation of a quantitative genetic model that describes the evolution of ecologically important phenotypic traits in heterogeneous environments (Via and Lande 1985, 1987; Via 1987). That work was not begun with the intention of modeling the evolution of the reaction norm. Instead, our goal was to produce a general description of the evolutionary dynamics of quantitative characters in a spatially variable environment. Having done so, however, it became clear that we had also described how both adaptive and nonadaptive phenotypic plasticity could evolve in populations that inhabit a spatial patchwork. There are now several additional quantitative genetic models that address the mechanisms of evolution of phenotypic plasticity and the reaction norm (Kirkpatrick and Heckman 1989; van Tienderen 1991; Gomulkiewicz and Kirkpatrick 1992). All of these models share a common premise—that stabilizing selection *within* each environment that is experienced by a given population drives characters toward an environment-specific phenotypic optimum. As populations evolve toward these (possibly different) optima in different environments, phenotypic plasticity results. As discussed below, it is the premise that selection occurs on trait values within environments rather than on change per se that conflicts most seriously with the idea that plasticity is a character in its own right that can evolve independently of trait means.

My purpose here is to review the logical implications of the idea that plasticity is a character in its own right in the light of what is currently understood about the genetics and evolution of quantitative characters. Some of these ideas are also considered in Via (1993). I will begin with a review of the basic biology involved in cases of phenotypic plasticity, consider how plasticity can be measured, and briefly discuss how adaptive phenotypic plasticity might evolve under natural selection. Then I will consider the following questions: (1) Is plasticity itself the target of selection? (2) Does phenotypic plasticity evolve separately from the mean of a trait?, and (3) Are there "genes for plasticity" and if so, what is their function?

What Is Phenotypic Plasticity?

Many characters of ecological importance exhibit phenotypic plasticity. Classical examples of plasticity tend to involve mostly dramatic and discontinuous phenotypic differences between individuals in different environments, such as differences in leaf shape when "amphibious" plant species develop in terrestrial or aquatic environments (Bradshaw 1965), as well as defensive structures induced by predators in invertebrates such as cladocera (Dodson 1989), rotifers (Gilbert 1966), or bryozoa (Harvell 1984, 1990). In addition, many behavioral, physiological, and biochemical traits also exhibit plasticity (examples in Hochachka and Somero 1984; Atkinson and Walden 1985). Although perhaps not as striking as the kind of switch between morphs that is often associated with phenotypic plasticity, many characters also change quantitatively if organisms are allowed to develop in different environments. Whether qualitative or quantitative, phenotypic changes across a set of environments define the norm of reaction.

When considering phenotypic plasticity, it is important to remember that while the proximal cues that trigger the phenotypic differences are environmental, the ability to respond to environmental cues is genetically based and the extent of that response can evolve under natural selection. Although the classic examples of plasticity listed here are generally presumed to be selectively advantageous, the mechanisms by which such adaptive responses to the environment evolve remain elusive.

The first step in understanding how plasticity can evolve is to determine how it can be measured. Two major types of phenotypic response to the environment (plasticity) have been considered in the literature. The first, and most commonly discussed, is the norm of reaction—the array of (possibly different) phenotypes that are produced by a single genotype across a range of environments. Central to the concept of the reaction norm is the idea that the environments involved are repeatable and predictable aspects of the organism's habitat, such as aquatic or terrestrial habitats for plants, or presence or absence of predators for the invertebrates that exhibit induced morphological defenses.

In contrast, a second type of plasticity has also been considered. This is the response to largely unpredictable variability within environments. In such unpredictable microenvironments, the generation of variable progeny, either by genetic means or through heightened sensitivity to small environmental fluctuations occurring during development (e.g., Bull 1987), can provide a selective advantage by ensuring that some progeny are suited to whatever the current environment might be (Bull 1987). When environmentally induced, such "noisy plasticity" is different from an adaptive reaction norm, because the result is variability in the phenotype within the current environment rather than the production of a particular phenotype. In other words, "noisy" plasticity is essentially shotgun variability that results from developmental instability, that is, hypersensitivity to small and effectively random environmental fluctuations (Bradshaw 1965; Bull 1987). In contrast, the phenotypic plasticity that leads to an adaptive reaction norm involves similar phenotypic change among all individuals in direct response to some predictable and repeatable environmental stimulus. Although this chapter will focus exclusively on the evolution of the norm of reaction, it is critical to keep these two types of plasticity separate, because they may evolve under different types of selection and they have been modeled in different ways (contrast Bull 1987 with Via and Lande 1985).

Measuring the Reaction Norm

Given any population of individuals, there will be some average response to a range of environments, defining the mean reaction norm. In a simple two-environment case, the difference in the mean phenotype in the two environments measures the average phenotypic plasticity. Around this mean reaction norm, there is likely to be some variability in reaction norms among genotypes. This genetic variation in plasticity is the raw material for the evolution of a change in the average reaction norm should changes in the environment produce selection toward a new optimum phenotype in one or more environments (Via and Lande 1985; Via 1987).

In order to estimate the mean and the genetic variation of the norm of reaction, it is necessary to test genotypes over a range of environments. The first step is either to generate groups of relatives (such as half- or full-sibs) using a standard quantitative genetic breeding design (e.g., Falconer 1989), or to assemble a group of clones if one is considering a parthenogenetic species. A group of progeny from each family or clone must then be tested in each of the environments of interest. A similar type of "split-brood" design can be used to test for local adaptation (see chapter 4 and Via 1984a, 1991).

The results of such an experiment can be analyzed using a two-way analysis of variance (ANOVA), with "Genotype" (i.e., family or clone) and "Environment" as the two factors. Although there are better ways to analyze the data from this type of experiment in order to obtain estimates of the parameters that influence the rate and form of reaction norm evolution (see below), interpretation of the ANOVA table can be a useful heuristic aid, particularly when only two environments are considered (table 3.1).

In particular, the main effect of "Environment" estimates the extent of the effect of the test environment or habitat on the phenotype of the average genotype. In a simple two-environment case, the "Environment" term estimates the mean phenotypic plasticity. The interpretation of the "Genotype" main effect can be complex, depending on how the F-test is constructed. It may either estimate variation among genotypes in the

TABLE 3.1

Interpretation of two-way ANOVA with respect to measurements of phenotypic plasticity in a simple two-environment case

Source	Meaning of F-Test
Genotype (V_G)	Genetic variation in overall mean phenotype (or, covariance of genotypic values across environments)?
Environment (V_E)	Effect of the environment on average phenotype? (mean phenotypic plasticity)
G × E ($V_{G \times E}$)	Variation among genotypes in response to environment? (genetic variation in phenotypic plasticity)
Error (V_e)	Variation among replicates within genotype-environment combinations
Total (V_P)	Total phenotypic variance

Notes: Terms in parentheses are the variance components estimable from each term. Note that the main effect for genotype may have two different interpretations, depending on the construction of the F-test (see text and Fry 1992).

"average" environment or estimate the covariance of genotypic values across environments (see Fry 1992). The latter interpretation provides a useful method for estimating the genetic covariance among character states (see below), but may have little meaning for multiple environments unless they are considered pairwise (Fry 1992). It should be noted that genetic variances within particular environments cannot be estimated from this overall "Genotype" main effect; these must be obtained from additional analyses in which observations from each environment are analyzed separately.

Finally, in this two-way ANOVA, genetic variation in response to the environment is estimated by the "Genotype × Environment" (G × E) interaction term. This interaction thus estimates genetic variation in phenotypic plasticity (see also Via 1987 for a discussion of this point). In the two-environment case, or if one takes the environments pairwise, the G × E interaction also suggests whether the values of a given character expressed in the different environments (henceforth, "character states") are genetically independent: if there is no significant G × E, then there is complete genetic dependence between the character states expressed in each environment. This is important because if there is not at least partial genetic independence of the character states in the different environments, a change in the norm of reaction cannot evolve (Via and Lande 1985). Unfortunately, however, there is no monotonically increasing relationship between the degree of genetic independence across environments and the G × E interaction variance. Thus, the G × E term does not always capture the extent to which a new reaction norm can evolve. Below, I consider how to improve on the estimate of genetic independence across

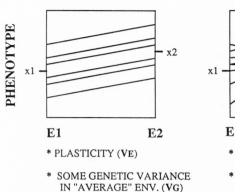

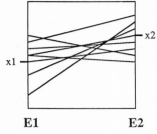

* PLASTICITY (VE)	* PLASTICITY (VE)
* SOME GENETIC VARIANCE IN "AVERAGE" ENV. (VG)	* NO GENETIC VARIANCE IN "AVERAGE" ENVIRONMENT (NO VG)
* NO VARIANCE IN PLASTICITY (NO VGxE)	* SOME GENETIC VARIANCE IN PLASTICITY (VGxE)

FIG. 3.1. Genotypic reaction norms in a simple two-environment case. For each geno-type, the mean value of the character state expressed in Environment 1 (E1) is plotted on the left vertical axis, and the mean value of the character state expressed in Environment 2 (E2) is plotted on the right vertical axis. The "slope" of the line connecting the means for each genotype gives a visual impression of that genotype's phenotypic plasticity (note that these are not regression lines). In both examples, there is some average phenotypic plasticity, which is seen as the difference in the means of all geno-types in the two environments ($|\bar{x}1 - \bar{x}2|$ on the figure, variance attributable to this is estimated by V_E from table 3.1). There is also roughly equivalent genetic variation within environments in both examples. *Left*: All genotypes respond similarly to the environment and have the same plasticity. Although there is "overall" genetic variation (V_G from table 3.1), there is no variation among genotypes in response to the environment (no V_{GxE} from table 3.1). *Right*: Genotypes differ in their response to the environment. This genetic variation in plasticity is the genotype × environment inter-action, V_{GxE}, from table 3.1.

environments using the genetic correlation between character states ex-pressed in the different environments (e.g., "body size in environment 1" and "body size in environment 2").

The meaning of the terms in the two-way ANOVA can be visualized graphically by plotting possible outcomes of a simple experiment that involves only two environments (fig. 3.1). In these plots, the mean pheno-typic values for each family or clone in each environment are shown on the vertical axes and are connected by the lines. Note that these are not regression lines, they are simple reaction norms that indicate graphically the difference in the values of the character states expressed by a genotype in the two environments. The "slope" of the lines thus indicates the extent of phenotypic plasticity for each genotype.

In both plots in figure 3.1, the mean phenotype is larger in Environ-ment 2 (E2) than it is in E1. Thus, in both examples, there is phenotypic plasticity at the level of the population mean. There is also some variability among genotypes within each environment in both examples. However, in the example in figure 3.1 (left), all genotypes respond similarly to the environment. The genotypic reaction norms are parallel, and there is no genetic variation in plasticity (no $G \times E$). There appears to be, however, some "overall" genetic variation that would be tested for significance by the main effect of "Genotype" in the ANOVA. One interpretation of this "overall " genetic variation is as variation among genotypic values averaged over environments. This "overall" genetic variance can be considered as variation around the "grand mean" of the genotypic values in the "average" environment. However, because individuals do not experience and are not selected in the average environment, the grand mean and the "overall" genetic variation are probably less relevant to evolution in a heterogeneous environment than are the means and genetic variances *within* environments. Furthermore, Fry (1992) notes that the variation in the "average" environment estimated by the "Genotype" main effect in a two-way ANOVA may have little relevance to genetic variance under field conditions, because it is highly dependent on the frequencies of the different environmental types, which are almost certain to differ in the field from those used in a designed experiment. One can also see how Fry's alternative interpretation of the "Genotype" main effect, the genetic covariance across environments (Fry 1992), comes about: in figure 3.1 (left) there is very high genetic covariance across environments due to the similarity of responses to the environment by the different genotypes.

In contrast to the example in figure 3.1 (left), genotypes in figure 3.1 (right) differ in both the magnitude and the direction of their phenotypic response to the environment. This variation would be estimated by the $G \times E$ variance in the ANOVA, and it corresponds to genetic variation in reaction norms, that is, to genetic variation in phenotypic plasticity. In the example in figure 3.1 (right), a significant main effect of "Genotype" would be unlikely, regardless of the mode of calculation (e.g., Fry 1992): (a) differences in genotypic values in the two environments would tend to cancel out, reducing genetic variation in the "average" environment; and (b) the lack of correspondence of genotypic values in the two environments would be likely to correspond to a low and nonsignificant genetic covariance across environments.

It is important to note that although this simple two-way ANOVA for the two-environment case provides a way of testing for a difference in the mean phenotype across environments and for visualizing genetic variation in plasticity through the $G \times E$ interaction, there are some drawbacks. Via and Lande (1985) showed that the rate and direction of reaction norm evolution is determined by the genetic variances within environments, in combination with the degree of independence of the character states expressed in the different environments (model described

below). There are two problems with the two-way ANOVA. First, no information is provided by this analysis on the magnitude of genetic variability
available for a response to selection *within* environments: this must be
estimated from separate ANOVAS on observations in each environment.
Second, a significant Genotype × Environment interaction only suggests
the potential for reaction norm evolution. As I will describe below, very
high values of the G × E term in a two-way ANOVA can correspond to a
lack of independence among the character states expressed in different
environments. Thus, the evolvability of the reaction norm is not a monotonically increasing function of the G × E variance, and estimates of the
genetic covariances of character states expressed in different environments are therefore preferred to estimates of the G × E variance (obtain-
able from the "Genotype" main effect in the ANOVA as described by Fry
1992, or using methods discussed by Via 1984b). If the environments of
interest are continuous rather than discrete, a covariance function can be
estimated that replaces the need for a matrix of pairwise genetic covariances across environments (methods in Gomulkiewicz and Kirkpatrick
1992).

How Does Selection Act on Phenotypic Plasticity?

In this section, I will focus on how selection acts on reaction norms, since
this may differ from patterns of selection on "noisy" plasticity (e.g., Bull
1987). Both spatial and temporal variation in the environment have
been considered to date in quantitative genetic models of reaction norm
evolution:

    1. *Spatial variation: each individual experiences a single environment in the
patchwork.* This scenario corresponds to Levins's (1968) "coarse grained" environment, in which individuals complete development in only one of the patch
types. This is the general situation considered by Via and Lande (1985, 1987),
Via (1987), van Tienderen (1991), and Gomulkiewicz and Kirkpatrick (1992).

    2. *Temporal variation between generations.* In this case, all individuals ex-
perience the same environment in a given generation, which may or may not
differ in the next generation (Gomulkiewicz and Kirkpatrick 1992).

    3. *Temporal variation within generations.* This is formally equivalent to se-
lection in a spatially variable environment when individuals move among
patches (Levins's [1968] "fine-grained environments"). When individuals can
experience several patch types, the evolutionary outcomes depend on whether
the traits of interest are labile, that is, subject to change during the lifetime, or
nonlabile and therefore unable to be affected by changes in the environment
after some critical period (Gomulkiewicz and Kirkpatrick 1992).

Though the details of the evolutionary trajectories followed by populations under each of these selection scenarios differ, as can the equilibrium reaction norms, they have three important features in common: (1) Selection is assumed to act only on the phenotypic character states expressed in the environment in which an individual finds itself at the moment; (2) within each environment, selection acts to move the population mean toward the optimum phenotype for that environment under stabilizing selection; and (3) evolution in variable environments requires that populations respond to selection on several character states in a "quasi-simultaneous" way, as parts of the population experience each environment or as the population experiences different environments in a sequence. Thus, each part of the reaction norm becomes adjusted in the environment in which it is expressed, and various parts of the norm are selected either simultaneously in different parts of the population (spatial variation) or sequentially (temporal variation). In all models, the optimum joint phenotype—and thus the optimal reaction norm—will not be attained unless there is sufficient genetic variation for all the character states and their combinations (Via and Lande 1985, 1987; Gomulkiewicz and Kirkpatrick 1992). In a few of the models, the optimum phenotype cannot be attained. For example, this occurs when there is a cost to plasticity (van Tienderen 1991) or when there is within-generation fluctuating selection on nonlabile traits (Gomulkiewicz and Kirkpatrick 1992). In these two models, the population evolves until it arrives at the best compromise among different selective forces.

Evolution of the mean phenotype in each environment toward an optimum phenotype requires that the character states expressed in the different environments be at least partially genetically uncorrelated with one another. If character states are genetically correlated, then an evolutionary change in the mean phenotype in one environment will lead to a correlated change in the mean phenotype of character states expressed in other environments. These correlated changes may take the population well away from the phenotypic optima in other environments, causing maladaptation of parts of the reaction norm. Depending on the magnitude of the cross-environment genetic correlations and the strength of selection, such maladaptive evolution may be quite temporary or of long duration (see Via and Lande 1985, Via 1987, and Gomulkiewicz and Kirkpatrick 1992 for more technical descriptions).

The importance of genetic independence between the character states to the evolution of plasticity can be visualized in figure 3.2. Within each environment, we can imagine selection acting to favor individuals near an optimum under stabilizing selection (noted by ⊗ on fig. 3.2). The arrows in the figure represent selection toward a larger optimum value of the phenotype, due to a change in ecological conditions within Environment

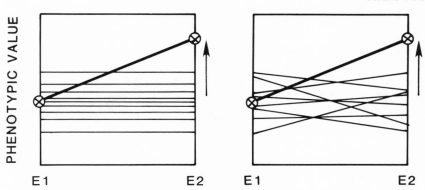

FIG. 3.2. Two hypothetical populations with different amounts of genetic variation in phenotypic plasticity. Genetic variation within environments is the same for all cases. Thin lines are reaction norms for individual genotypes. If ecological changes in Environment 2 changed the optimal phenotype under stabilizing selection upward, then selection (indicated by the arrows) would attempt to push the population mean toward the new optimum (indicated by ⊗). This new force of selection favors a larger phenotype in E2, while the phenotype in E1 is selected to remain the same. At *left*, where there is no variation among genotypes in the norms of reaction, the new norm can never evolve. At *right*, genetic variation in reaction norms is present, so the population can eventually adjust the mean in E2 and evolve to a new level of phenotypic plasticity. If selection is weak, the genetic variance and the G × E interaction will remain roughly constant during this process (Via and Lande 1987). (Reprinted with permission from Via 1987)

2. In this case, attainment of the new optimal reaction norm would entail adjustment of the mean phenotype within the changed environment while holding the mean value of the other character state constant. Shifting the mean phenotype in one environment while holding it constant in the other requires that the character states be at least partially genetically independent. In most cases, this independence is manifest as Genotype × Environment interaction, or nonparallel reaction norms. In the example shown in figure 3.2, only the population in the graph on the right could respond adaptively to a new force of selection to change the mean in Environment 2. In the graph on the left, the population would show a maladaptive correlated increase in the phenotype in Environment 1, and would eventually reach equilibrium at a compromise between selection toward the optimum phenotype in each environment. Later I will discuss how the genetic independence of the character states expressed in different environments that is required for reaction norm evolution can be quantified more precisely using the genetic correlations among character states instead of the genotype × environment interaction.

In the view of evolution in heterogeneous environments embodied by the quantitative genetic models (e.g., Via and Lande 1985, 1987; Via

1987; van Tienderen 1991; Gomulkiewicz and Kirkpatrick 1992), the mean reaction norm evolves *as a byproduct* of stabilizing selection on the mean values of the character states expressed in each environment. The optimal reaction norm is simply the one that corresponds to the optimum phenotype in each environment.

Does Selection Ever Act on Plasticity Itself?

In the quantitative genetic models of the evolution of phenotypic plasticity (Via and Lande 1985, 1987; Via 1987; van Tienderen 1991; Gomulkiewicz and Kirkpatrick 1992), the reaction norm evolves as a byproduct of selection on character states expressed within environments. However, the view that plasticity is a character in its own right and can evolve independently of trait means (Schlichting and Levin 1984; Schlichting 1986; Scheiner and Lyman 1989, 1991) implies that selection may at least sometimes act directly on plasticity itself, that is, on the change in the phenotype across environments. Scheiner and Lyman (1989) define the "heritability of plasticity" as the proportion of the total phenotypic variance that is attributable to the genetic variance in plasticity (i.e., for a two-environment case analyzed by an ANOVA such as the one seen in table 3.1, the genotype × environment interaction variance divided by the total phenotypic variance). They write:

> Knowing the heritability is important if we are to make predictions about the response of a population to environmental variation. For example, suppose that in a complex environment there was equal selection for individuals to specialize on particular environments or to increase their plasticity. If the plasticity of a trait had a greater heritability than the trait itself we might expect the primary evolutionary response to be an increase in the plasticity of individuals. [p. 96]

Scheiner and Lyman (1989) evidently believe that the selective forces acting on the degree of phenotypic change across environments (plasticity) and those acting on trait values can be separate. What does this imply?

If selection were to act directly on plasticity, then two genotypes with the same degree of phenotypic change across environments would be expected to have the same fitness. Figure 3.3 shows reaction norms for three genotypes. Presuming stabilizing selection within environments, genotype 1 happens to have the optimum phenotype (indicated by the asterisk) in each of the two environments. Genotype 2 shows the same degree of plasticity, but produces a phenotype that is far from the optimum in both environments. How can these two genotypes ever have the same fitness? Similarly, the fitness of genotype 3, which shows less plasticity but has

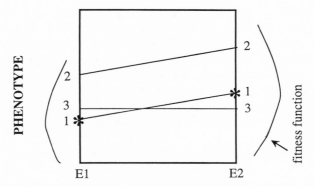

FIG. 3.3. Hypothetical reaction norms for three genotypes. Genotype 1 happens to produce the optimal phenotype under stabilizing selection in both environments (noted by the asterisk). Genotype 2 has the same plasticity, but is far from the optimum in both environments. Genotype 3 has a different plasticity, but is closer to both optima than is genotype 2. Under stabilizing selection within environments, genotypes 1 and 3 would have the highest fitnesses; the fitness of genotype 2 would be expected to be lower than that of the other two genotypes. If selection were to act on plasticity itself, genotypes 1 and 2 should have the same fitness, and the fitness of genotype 3 would be lower because it has a different plasticity.

character state values closer to the optimum in both environments, must have a higher average fitness than genotype 2, which is not favored in either environment.

If we assert that selection acts only on plasticity (the degree of change across environments) and not on trait values, then it appears that we must abandon the stabilizing selection model. However, genotypes 1 and 2 would also not be expected to have the same fitness under directional or disruptive selection within environments. Thus, the assertion that selection acts on plasticity itself is not compatible with any model of selection on the characters within environments.

Scheiner and Lyman (1991) justify their assertion that reaction norms evolve because selection acts on plasticity itself by using an argument about the selective advantage of variable progeny, that is, an argument about the advantages of "noisy plasticity" (p. 105). However, the benefits accorded to the production of phenotypically variable progeny through "noisy plasticity" are likely to be very different from the selective advantage afforded a genotype by having a particular set of phenotypes across a range of environments (Bull 1987).

If selection acts on trait values within environments and the norm of reaction evolves as a by-product of this selection, then estimation of the heritability of plasticity serves more of a heuristic than a practical function in the study of reaction norm evolution. It is certainly useful to know that genetic variation in plasticity exists, because this suggests that semi-independent evolution of trait values in different environments can occur. However, estimates of the heritability of plasticity are limited by the lack of direct correspondence between G × E variance and the evolvability of the reaction norm: the very highest values of the G × E term correspond to a lack of independence of character states across environments (see next section and fig. 3.4). Moreover, using the heritability of plasticity in a prediction equation (such as $R = h^2 s$, e.g., Falconer 1989) is not appropriate if selection acts on trait values within environments rather than on plasticity itself.

Evolution of the Norm of Reaction

Via and Lande (1985) presented a quantitative genetic model of evolution in a spatially variable environment that describes how phenotypic plasticity can evolve by natural selection on trait values within environments. The key feature of this model is that values of a given character in different environments are regarded as potentially independent character states, which each have a measurable genetic variance and which may be genetically correlated with one another. This approach follows that taken by Falconer (1952), who first discussed how the Genotype × Environment interaction could be viewed as a genetic correlation between "character states" expressed in two environments (a discrete set of multiple environments can be considered by expanding the matrix of genetic variances and covariances among character states, e.g., Via 1987; for continuous environments a covariance function can be estimated and substituted for the covariance matrix, e.g., Gomulkiewicz and Kirkpatrick 1992). Using this approach, Via and Lande (1985) formalized the effects of different degrees of genetic independence of the character states expressed in different environments on the rate of reaction norm evolution.

The relationship between the Genotype × Environment interaction and the genetic correlation between character states for a simple two-environment case can be seen in figure 3.4. This diagram illustrates why the degree of genetic independence between character states in different environments can be more clearly estimated using the genetic correlation of character states than by using the Genotype × Environment interaction. In the left-hand panels, data are plotted as genotypic reaction norms, with each line connecting the means for a single genotype in two environments. In the right-hand panels, the same data are plotted as scatterplots of genotypic means in the two environments. These scatterplots illustrate the magnitude of the "cross-environment" genetic correlation for a given value of the Genotype × Environment interaction.

There is progressively more Genotype × Environment interaction going from top to bottom in figure 3.4. However, one can see that genetic inde-

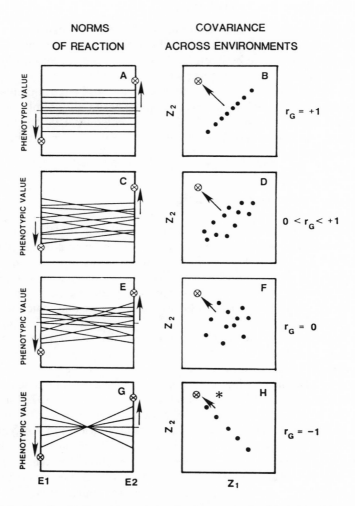

FIG. 3.4. Four levels of genetic variation in phenotypic plasticity plotted in two ways. *A, C, E, G*: Genotype × environment interaction plots. *B, D, F, H*: Scatter plots from which the cross-environment genetic correlations between character states (r_G) can be estimated. Each line or point corresponds to the mean phenotype that a given genotype expresses in each of two environments. Axes for the left-hand panels are as in figure 3.1. In the right-hand panels, z_1 is the value of the character state that is expressed in Environment 1, and z_2 is the phenotypic value of the character state that is expressed in Environment 2. In all plots, the mean phenotype expressed in each environment is the same, that is, the average level of phenotypic plasticity is initially zero. ⊗ indicates a new optimal level of phenotypic plasticity toward which we may imagine the population is being selected, with stabilizing selection on the two character states in the directions of the arrows. Genetic variation for a change in the current level of phenotypic plasticity is zero in A, B, G, and H (the cross-environment genetic correlation $r_G = \pm 1$), intermediate in C and D ($0 < r_G < 1$), and high in E and F ($r_G = 0$), even though the magnitude of the G × E interaction variance increases monotonically from top to bottom of the figure. New optima ⊗ or * will never be attained in B or H, respectively.

pendence of the character states in the two environments is at a maximum at an intermediate value of the $G \times E$ variance when the genetic correlation between the character states is equal to zero (fig. 3.4E,F). Greater magnitudes of $G \times E$ correspond to negative genetic correlations between character states expressed in the two environments and to progressively less genetic independence between the character states. A significant cross-environment genetic correlation of either sign indicates that the character states are not independent. Thus, any cross-environment genetic correlation of absolute value greater than zero suggests the possibility of genetic constraint on independent adjustment of the mean phenotype expressed within each of the different environments, and thus a constraint on the evolution of an adaptive reaction norm. Because the Genotype $\times$ Environment interaction variance continues to increase after the degree of genetic independence reaches its maximum at a cross-environment correlation of zero, it is less useful as a metric of selectable genetic variation in phenotypic plasticity than is the cross-environment genetic correlation.

Note that whether a given cross-environment genetic correlation will be constraining depends critically upon the way in which natural selection acts to change the mean phenotype expressed in different environments. In figure 3.4, for example, a positive genetic correlation is most constraining with respect to response to selection toward the hypothetical new optimum. However, had selection acted to change the means in both environments in the same direction, a negative genetic correlation would have been most constraining (see other examples in Via and Lande 1985; Via 1987).

The magnitude of the cross-environment genetic correlation can be interpreted in terms of an underlying genetic mechanism in much the same way as can the value of other genetic correlations (cf. Falconer 1989). Non-zero genetic correlations result when the same alleles (correlation through pleiotropy) or sets of alleles (correlation through linkage disequilibrium) affect the character states in two environments in the same way.

The lack of genetic correlation among character states in different environments may be attributable either to variation in the environmental sensitivity of alleles at particular loci, or to loci with environment-specific effects that affect the phenotype in only a single environment. Such loci introduce genetic variation in the character state expressed in that particular environment that is uncorrelated with genetic variation in character states that are expressed in other environments. This uncorrelated genetic variation facilitates the independent adjustment of character state means in each environment, and thus speeds the evolution of an adaptive reaction norm.

Selection toward a hypothetical new reaction norm is indicated in figure 3.4 either by arrows up in E1 and down in E2 (for the G × E plots), or by an arrow toward the asterisk (for the scatterplots). Via and Lande (1985) showed that response to this joint force of selection as reflected in changes in the mean values of the character states expressed in each of the two environments ($\Delta \bar{z}$), is influenced both by the genetic variances of the character states within environments (Gii), and by the genetic covariance between the states (Gij). The response to selection of the character state expressed in environment 1 is thus:

$$\Delta \bar{z}_1 = G_{11} q_1 \, ln \bar{W}_1 + G_{12} q_2 \, ln \bar{W}_2$$

response = direct response + indirect (correlated) response,   (3.1)

where q_i is the fraction of the population selected in the ith environment in a soft selection scenario, and $ln \bar{W}_1$ is the magnitude of the direct force of selection on the character state expressed in the ith environment (see Via and Lande 1985and Via, 1987 for technical details). As seen in eq. 3.1, the response to selection in a given environment involves both a direct response to selection in that environment plus an indirect or correlated response to selection in the other environment. The magnitude of the correlated response is directly proportional to the genetic covariance between the character states expressed in the two environments.

For the two cases of perfect genetic correlation between character states expressed in two environments (fig. 4A,B,G,H), most new norms can never be reached because there is no independent genetic variation in the two environments. In such a case, the joint phenotype can never evolve off the line that can be drawn through the bivariate genetic values. For the other panels, in which there is some independent variance in each environment, evolution of a new reaction norm will proceed at a greater or lesser pace, depending on the magnitude and sign of the cross-environment genetic correlation. Via and Lande (1985) and Via (1987) present several examples that allow the effects of the cross-environment genetic correlation on the evolution of a new reaction norm to be visualized graphically.

This model has not been without criticism. In particular, it has been criticized because there is no provision for plasticity to evolve separately from trait means: "Treating plasticity as a trait in its own right allows for possibilities not considered by Via and Lande. For example, a trait can evolve as a correlated response to selection on the plasticity of that trait, or there can be joint evolution on the plasticities of different traits" (Scheiner and Lyman 1989, p. 105). Also, "genes for plasticity" were not invoked in our models, and this was criticized: "Genes for plasticity of a trait may be separate from those responsible for the expression of that trait. Thus, they may evolve separately, and models need to incorporate these possibilities" (Scheiner and Lyman 1989, p. 105). In the following sections, I will consider whether plasticity is likely to evolve separately from trait means, and will explore the implications of invoking "genes for plasticity."

Does Plasticity Evolve Separately from the Mean of a Trait?

The idea that plasticity might evolve separately from trait means dates back at least to the classic review of Bradshaw (1965), who reviewed many cases of phenotypic plasticity in plants. Based on his compilation of examples, Bradshaw emphasized that plasticity is not a property of the entire genome, but is "a property specific to individual characters in relation to specific environmental influences" (p. 119). He also noted that closely related species, and even varieties within species, can differ dramatically in their sensitivity to the environment. From this he concluded that "such differences (in plasticity among related species) are difficult to explain unless it is assumed that the plasticity of a character is an independent property of that character and is under its own specific genetic control" (p. 122).

Although this seems like a reasonable conclusion to draw from his observations, Bradshaw assumed a mechanism in the absence of any genetic model of evolution. In 1965 there were no population genetic models for the evolution of quantitative traits under natural selection, even in single environments. Thus Bradshaw had little opportunity to evaluate his proposed mechanism in a theoretical context that could reveal its potential for contributing to the process of the evolution of adaptive phenotypic plasticity. Now that several genetic models for the evolution of the reaction norm have been proposed (Via and Lande 1985; van Tienderen 1991; Gomulkiewicz and Kirkpatrick 1992), the observations made by Bradshaw can be seen to be consistent with explanations other than the independent evolution of plasticities and trait means. For example, perhaps species with differing plasticities have been under selection in a different array of environments, and thus have had the trait means within environments adjusted in different ways. It is no longer necessary to suppose that plasticities and trait means evolve independently by virtue of separate genetic control to explain the differences among species that Bradshaw observed.

However, Bradshaw's review continues to be very influential (cf. Schlichting 1986), and perhaps because of this influence, the idea that plasticities and trait means evolve independently is now quite popular.

The logic behind the current expression of this idea can be stated as follows. Because plasticity is considered by proponents of this view to be a character in its own right (Bradshaw 1965), it can evolve independently of the mean of the trait (Schlichting and Levin 1986; Schlichting 1986; Jinks and Pooni 1988; Scheiner and Lyman 1989, 1991). However, which trait mean is under consideration is sometimes unclear: is it the mean phenotype within an environment, or the grand mean over all environments?

Data cited by Schlichting (1986) to support the independence of plasticity illustrate that the amount of change across environments is independent of the grand mean over all environments. It is clear that very different sets of means in different environments could be averaged to yield the same grand mean over all environments. However, is the grand mean important in the process of reaction norm evolution? Individuals do not experience the "average" environment; instead, selection occurs within environments. Thus I contend that the grand mean phenotype probably does not evolve under direct selection, but is, like the reaction norm, a by-product of selection on trait means within environments. Because the reaction norm is defined by the differences among mean phenotypes in different environments, it is difficult to imagine how plasticity could be independent of the within-environment means.

Scheiner and Lyman (1991) used a selection experiment to test the hypothesis that plasticity and trait means are independent with a selection experiment. They selected lines of *Drosophila melanogaster* for thorax size at 19°C and 25°C and also for the plasticity of thorax size. They found that they could increase the difference in size between flies at 19°C and 25°C (their definition of plasticity) the most by selecting up at 19°C and down at 25°C (Scheiner, pers. comm.). Thus their "selection for plasticity" was de facto selection on trait means, and so it is not surprising that they observed correlated responses to selection between plasticity and trait means.

Are There "Genes for Plasticity"?

The idea that there are "genes for plasticity" arises fairly directly from Bradshaw's (1965) assertion that plasticity is a character in its own right and is under separate genetic control. Loci for plasticity presumably effect a change in the phenotype across environments without influencing trait means. Schmalhausen (1949) proposed that alleles that influence trait values can have different environmental sensitivities, and asserted that the variable expression of mutant alleles in different environments is so pervasive that it provides a continuous source of raw material for the evolution of new reaction norms.

However, this is not the implication of the "genes for plasticity" idea. Scheiner and Lyman (1991) stress the existence of separate sets of genes and discuss the hypothesis that "plasticity is due to genes that determine the magnitude of response to environmental effects which interact with genes that determine the average expression of the character" (p. 25). Their conception of the genetic basis of plasticity seems to ignore variation of the type documented by Schmalhausen (1949), in which single mutant alleles at the same locus affect a character differently in different environments. For example, the allele "pennant" causes flies to produce normal wings at low temperature and small wings at high temperature, while another allele at the same locus, "vestigial," has just the opposite effect (Schmalhausen 1949). In Scheiner and Lyman's terminology, is this locus a gene for plasticity or a gene for the character?

Scheiner and Lyman (1991) also propose a model for the action of plasticity genes. Their model entails three classes of loci: (1) loci that influence trait means; (2) loci that influence whether an individual will respond to an environmental cue; and (3) loci that determine the magnitude of such a response. One example of such a system might involve a drug or hormone receptor (Scheiner, pers. comm.). In the environment containing the drug, only those individuals with appropriate alleles at the receptor locus could respond to the compound, and the amount of the response might be influenced by loci that affect drug or hormone metabolism. These effects then modify the values of character states that are expressed in the environment(s) containing the compound.

However, such loci can also be interpreted as genes with an environmentally specific expression: trait values are only affected by such a locus in the environment that contains the drug or hormone. In other environments, variation is not affected by the receptor locus. As discussed above, loci with environmentally specific expression are already incorporated in current quantitative genetic models of the evolution of reaction norms. To be a true "plasticity gene," it would seem that a locus must produce change across environments in a way that is not dependent on the particular environments concerned. In my opinion, if loci for the general ability to learn exist, they might provide a good example of genes for behavioral plasticity. However, the key to the selective advantages of such genes is the flexibility of response that they might permit in novel environments. In contrast, reaction norm evolution concerns responses to predictably occurring environments. Thus, even if there are "learning genes," they may not provide the appropriate model for the evolution of reaction norms (in contrast to assertions by Schlichting and Pigliucci 1993).

Recently, Schlichting and Pigliucci (1993) have proposed a different definition of "plasticity genes": "We define plasticity genes as regulatory loci that exert environmentally-dependent control over structural gene expression, and thus produce a plastic response. The evidence for such genes requires that we reevaluate models of reaction norm evolution" (Schlichting and Pigliucci 1993). I fully agree that such regulatory loci may be involved in controlling the expression of the loci that influence trait values in particular environments. However, in contrast to the assertion made by these authors, the role of regulatory loci does not demand that quantitative genetic models need to be replaced. Quantitative genetics models are blind to the precise genetic mechanisms that produce the genetic variances and covariances among traits within environments. As long as alternative alleles at the regulatory loci each affect several "structural" loci, then assumptions of normality of breeding values within environments will not be violated. Thus, the regulatory loci act "behind the scenes" to influence the expression of the loci that produce the phenotype in a given environment, and they are selected only indirectly through the phenotypic variation that they influence. For these reasons, spelling out a role for regulatory loci is completely consistent with the more abstract quantitative genetics models, which describe only the consequences, not the causes, of genetic covariances across environments.

Summary

Several recent studies have proposed that phenotypic plasticity evolves independently of trait means, and that separate loci influence trait values and plasticities. These ideas conflict with current quantitative genetic models of the evolution of phenotypic plasticity, which all assume that characters are under stabilizing selection within environments. In these models, in the absence of costs, each of the character states that collectively comprise the reaction norm is subject to selection only in the environment in which it is expressed.

Quantitative genetic models suggest that adaptive reaction norms evolve not by selection on plasticity itself, but as a by-product of selection on trait means within environments. In this sense, the reaction norm is an emergent property rather than a character in its own right. Results from the quantitative genetic models indicate that the evolution of adaptive reaction norms requires partial genetic independence of character states expressed in different environments. This genetic independence can be achieved either if some loci have environmentally specific effects, or if alleles at the same loci differ in environmental sensitivity. Although phenotypic plasticity may be independent of the grand phenotypic mean over all environments, it is unlikely to be independent of trait means within environments. Finally, invoking "genes for plasticity" is unnecessary because current quantitative genetic models already incorporate the possibility of environment-specific gene expression. Moreover, such models concern the effects of patterns of genetic variances and covariances among character states in different environments rather than the precise genetic mechanisms that produce the covariance patterns.

Acknowledgements

I am very grateful to Russ Lande for his help in formulating the models that provide the basis for these thoughts on phenotypic plasticity; to Sam Scheiner for explaining his 1991 paper to me; to Richard Gomulkiewicz and Mark Kirkpatrick for sending me a prepublication copy of their manuscript on reaction norm evolution; and to Carl Schlichting, Richard Gomulkiewicz, and Russ Lande for illuminating discussions. The preparation of this paper was supported by USDA Competitive Grant # 88-37513-3747.

References

Atkinson, B. G., and D. B. Walden. 1985. *Changes in Eukaryotic Gene Expression in Response to Environmental Stress.* Academic Press, London.

Bradshaw, A. D. 1965. Evolutionary significance of phenotypic plasticity in plants. *Advances in Genetics* 13: 115–155.

Bull, J. J. 1987. Evolution of phenotypic variance. *Evolution* 41: 303–315.

Dodson, S. 1989. Predator-induced reaction norms. *BioScience* 39: 447–452.

Falconer, D. S. 1952. The problem of environment and selection. *Amer. Nat.* 86: 293–298.

Falconer, D. S. 1989. *Introduction to Quantitative Genetics.* 3d ed. Longman, New York.

Fry, J. D. 1992. The mixed model analysis of variance applied to quantitative genetics: Biological meaning of the parameters. *Evolution* 46: 540–550.

Gause, G. F. 1947. Problems of evolution. *Trans. Connecticut Acad. Sci.* 37: 17–68.

Gilbert, J. J. 1966. Rotifer ecology and embryological induction. *Science* 151: 1234–1236.

Gomulkiewicz, R., and M. Kirkpatrick. 1992. Quantitative genetics and the evolution of reaction norms. *Evolution* 46: 390–411.

Harvell, C. D. 1984. Predator-induced defense in a marine bryozoan. *Science* 224: 1357–1359.

Harvell, C. D. 1990. The ecology and evolution of inducible defenses. *Quart. Rev. Biol.* 65: 323–340.

Hochachka, P. W., and G. N. Somero. 1984. *Biochemical Adaptation*. Princeton University Press, Princeton, N.J.

Jinks, J. L., and H. S. Pooni. 1988. The genetic basis of environmental sensitivity. In *Proceedings of the Second International Conference on Quantitative Genetics*, ed. B. S. Weir, E. J. Eisen, M. M. Goodman, and G. Namkoong, pp. 505–522. Sinauer, Sunderland, Mass.

Kirkpatrick, M., and N. Heckman. 1989. A quantitative genetic model for growth, shape, and other infinite-dimensional characters. *J. Math. Biol.* 27: 429–450.

Levins, R. 1968. *Evolution in Changing Environments*. Princeton University Press, Princeton, N.J.

Scheiner, S. M., and R. F. Lyman. 1989. The genetics of phenotypic plasticity. I. Heritability. *J. Evol. Biol.* 2: 95–107.

Scheiner, S. M., and R. F. Lyman. 1991. The genetics of phenotypic plasticity. II. Response to selection. *J. Evol. Biol.* 3: 23–50.

Schlichting, C. D. 1986. The evolution of phenotypic plasticity in plants. *Ann. Rev. Ecol. Syst.* 17: 667–693.

Schlichting, C. D. 1989. Phenotypic plasticity in *Phlox*. II. Plasticity of character correlations. *Oecologia* 78: 496–501.

Schlichting, C. D., and Pigliucci, M. 1993. Control of phenotypic plasticity via regulatory genes. *Amer. Nat.* (in press).

Schlichting, C. D., and D. A. Levin. 1984. Phenotypic plasticity in annual *Phlox*: Tests of some hypotheses. *Amer. Jour. Bot.* 71: 252–260.

Schlichting, C. D., and D. A. Levin. 1986. Phenotypic plasticity: An evolving plant character. *Biol. J. Linnean Soc.* 29: 37–47.

Schmalhausen, I. I. 1949. *Factors of Evolution: The Theory of Stabilizing Selection*. University of Chicago Press, Chicago.

Stearns, S. C. 1989. The evolutionary significance of phenotypic plasticity. *BioScience* 39: 436–445.

van Tienderen, P. H. 1991. Evolution of generalists and specialists in spatially heterogeneous environments. *Evolution* 45: 1317–1331.

Via, S. 1984a. The quantitative genetics of polyphagy in an insect herbivore. I. Genotype-environment interaction in larval performance on different host plant species. *Evolution* 38: 881–895.

Via, S. 1984b. The quantitative genetics of polyphagy in an insect herbivore. II. Genetic correlations in larval performance within and across host plants. *Evolution* 38: 896–905.

Via, S. 1987. Genetic constraints on the evolution of phenotypic plasticity. In *Genetic Constraints on Adaptive Evolution*, ed. V. Loeschcke. Springer-Verlag, Berlin.

Via, S. 1991. The genetic structure of host plant adaptation in a spatial patchwork: Demographic variability among reciprocally transplanted pea aphid clones. *Evolution* 45: 827–852.

Via, S. 1993. Adaptive phenotypic plasticity: Target or byproduct of selection in a variable environment? *Amer. Nat.* (in press).

Via, S., and R. Lande. 1985. Genotype-environment interaction and the evolution of phenotypic plasticity. *Evolution* 39: 505–523.

Via, S., and R. Lande. 1987. Evolution of genetic variability in a spatially variable environment: Effects of genotype-environment interaction. *Genet. Res.* 49: 147–156.

Woltereck, R. 1909. Weitere experimentelle Untersuchungen über Artveränderung, speziell über das Wesen quantitätiver Artenunterschiede bei Daphniden. *Verh. D. Zool. Ges.* 1909: 110–172.

4

Population Structure and Local Adaptation in a Clonal Herbivore

Introduction

Evolution is more than just a set of events that occurred in the past to produce the array of species that currently inhabits our planet. Evolution also occurs in the present as the genetic composition of contemporary populations tracks changes in the environment caused by natural or man-made disturbance. Darwin recognized that if two conditions are satisfied, then evolution will occur. These two conditions are that (1) phenotypic differences among individuals cause them to make a greater or lesser contribution to the next generation (natural selection), and that (2) these phenotypic differences are inherited (genetic variation).

Darwin's two conditions essentially outline a program for studying evolution as an ongoing process in contemporary populations. Such a program involves experimentally estimating the magnitude of genetic variation in ecologically important characters and determining the magnitude of natural selection they may have experienced in the past or to which they may currently be subject. By interpreting such estimates in the light of genetic models of evolution, we not only obtain some insight into causes of present-day differences among populations and/or species (Lande 1979; Charlesworth et al. 1982; Schluter 1984; Turelli, Gillespie, and Lande 1989; Via 1991a), but we can determine likely rates and directions of future evolutionary change (e.g., Lande 1979; Conner and Via 1992).

The study of the genetic population structure of ecologically important traits in species that inhabit different environments is one step in the application of this approach. Population structure refers to the organization of genetic variability within and between populations. We can use these patterns of variation to infer the extent of particular evolutionary forces acting on populations. When genetic structure is estimated using protein electrophoretic data or other neutral markers, valuable insight can be obtained into the extent to which genetic drift, population subdivision, and inbreeding influence patterns of variability within and between populations (e.g., Loveless and Hamrick 1984). Population structure at neutral loci can also reveal the extent of gene flow between populations (Slatkin 1987; see also chapter 1, this volume). However, analyses of variation at neutral loci are less useful for understanding patterns of variation in ecologically important quantitative traits because natural selection often causes the population structure of such characters to differ from that estimated for loci that evolve primarily by genetic drift. If one is specifically interested in the extent to which varying patterns of selection has led to population differentiation and/or local adaptation, then methods that focus on estimating genetic variation within and between populations in the phenotypic characters themselves must be employed.

Evaluating genetic population structure for quantitative characters involves determining the extent of genetically based phenotypic differentiation within and between populations. Differences between populations from different environments can be used to make inferences both about the extent of natural selection in the past and about the selective consequences of present-day migration. Estimates of genetic variability within populations permit us to study the process of evolution and to evaluate likely or unlikely patterns of future evolution in ecologically important characters. By studying local populations in which migration between environments can potentially occur and which are large enough to minimize genetic drift, one can reasonably hypothesize that population differentiation is due to selective differences among habitats. This hypothesis can be tested experimentally by comparing the performance of individuals originating from different habitats in a reciprocal transplant.

When assessing the genetic structure of populations with respect to performance in different habitats, it is useful to have a comprehensive measure of lifetime fitness. This can be obtained by measuring demographic characters. When there is genetic variation in demographic characters such as age-specific survival or fecundity, then genotypes can be expected to contribute differentially to population growth, in other words, to differ in fitness (Charlesworth 1980; Lande 1982). Thus genotype-specific life tables can be used to estimate fitness and to study natural selection among genotypes. In the study of population structure, estimates of genotypic demographies in the different environments can reveal not only the genetic structure of demographic characters, but also something about selective differences among environments.

In this chapter, I will summarize the results of a series of experiments that were designed to reveal the genetic population structure of characters associated with host plant use in a polyphagous herbivore, the pea aphid (*Acyrthosiphon pisum*). By studying the genetic structure of host adaptation in aphid populations that inhabit a spatial patchwork of host plants, inferences can be made about the extent of local adaptation of subpopulations in closely adjacent habitats, the strength of selection that might produce such population differentiation, and the genetic potential for further evolution of host adaptation in pea aphid populations. The field experiments to estimate genetic population structure were followed by laboratory experiments designed to show the extent to which specialized performance on the two hosts could be modified by experience and to investigate the genetic architecture of resource use specialization. In this research program, field and laboratory experiments were combined in an attempt to balance the realism offered by experiments in natural settings with the precision and control available in the laboratory environment.

The Genetics of Ecologically Important Phenotypic Characters

To study the genetic structure of ecologically important characters in pea aphid populations, we must be able to measure genetic variability in traits such as demography and life history, morphology, behavior, disease resistance, and parasitoid or predator avoidance. The techniques of quantitative genetics provide a way to estimate patterns of genetic variability in phenotypic characters such as these that are likely to be influenced by many loci. These methods, originally derived by R. A. Fisher and Sewall Wright, use phenotypic resemblances of relatives to estimate genetic variation (review in Falconer 1989). Although only the genetic component of the phenotypic variation in a character influences its response to selection, the genetic values of individuals cannot be directly observed because the phenotype is a composite of genetic and environmental influences. Thus the essential problem of quantitative genetics is to partition phenotypic variation (which can be observed) into its genetic and environmental components (which cannot be directly observed).

Although the statistical analysis of quantitative traits can be somewhat technical, the basic logic of the method is fairly simple: related individuals resemble one another because they share genes. For example, by knowing the degree of genetic relatedness between individuals such as parents and offspring, the mean phenotypic value of the offspring can be used to estimate the genetic value of the parents. Variance in these genetic values is the estimated genetic variance in the character. For a complete outline of the theory that relates phenotypic resemblances among relatives to the underlying genetic variability, see Falconer (1989).

Quantitative genetics is an exceptionally useful tool for the ecologist interested in evolution because these methods permit the genetic portion of phenotypic variability to be estimated for virtually any phenotypic character. Furthermore, by sampling individuals for genetic analyses from different habitats and keeping track of the site of origin for each individual, we can partition genetic variability in any of the phenotypic traits of interest into variation within populations and genetically based differences between populations. When genotypes from each habitat are reciprocally transplanted so that some replicates are placed in each habitat, their relative performance in each environment can be measured. Reciprocal transplantation is thus an important tool in the analysis of genetic population structure for quantitative traits (Briggs and Walters 1984, 308–315; Via 1990, 1991).

Demographic Estimates of Lifetime Fitness

The objective of this analysis of population structure in pea aphids was to determine the extent of genetic variability within and between aphid populations in performance on two different host plants. From such an analysis, we can determine the extent to which the host plants are actually different selective environments for the herbivores and measure both the current magnitude of local adaptation and the potential for further evolution of specialization.

In evaluating the extent of local adaptation, it is useful to have a measure of fitness that accurately estimates the ability of each genotype to contribute to the next generation. For populations with overlapping generations, it is necessary to take both the timing of reproduction and its magnitude into account to predict population growth (review in Wilson and Bossert 1971). Because fitness is the relative contribution of an individual to population growth, estimates of fitness in such populations must therefore also combine the timing and magnitude of reproduction. Simple measures of life-history components such as longevity or total fecundity are not always well correlated with contribution to the next generation (Travis and Henrich 1986). However, such measures can be placed into the context of population growth using demographic concepts of fitness (Charlesworth 1980; Lande 1982; Lenski and Service 1982; Travis and Henrich 1986; van Groenendael et al. 1988).

Charlesworth (1980) showed that for populations of noninteracting clones growing in a density-independent way, the clone-specific rate of population increase is a good estimate of clonal fitness because it closely estimates the relative contributions that clones with different life histories will make to the next generation. Charlesworth also showed how differential rates of population increase for clones (differential fitness) lead to clonal selection: clones with the most rapid rate of increase in a given environment become numerically dominant in the clonal assemblage, while clones with other life histories become displaced.

During the summer season in which reproduction is strictly parthenogenetic, pea aphids conform to the simple dynamics that Charlesworth (1980) modeled. During the clonal phase, no recombination occurs, so each clone essentially grows independently within the population at a rate determined by its particular demography. The population is thus a composite of these independently growing clones, and the population growth rate is a weighted average of the clonal growth rates, as Charlesworth describes (1980, pp. 65–69). It is likely that clonal population growth in pea aphid populations can be described by a density-independent model because the large mortality that occurs in alfalfa and clover fields when the farmers cut the fields (every forty-five days) effectively keeps the populations at low densities. Field sampling supports this hypothesis: population numbers of pea aphids appear to increase roughly exponentially between cuts (K. Hural, unpublished data).

In the spring, clonal diversity in pea aphid populations is expected to be at its peak because of the genetic variation released by recombination the previous fall (Lynch and Gabriel 1983; Lynch, chapter 6, this volume). If clones vary in fitness, then selection among clones will occur by a process of clonal replacement over the summer as the more successful clones become numerically dominant. Therefore, if we are interested in whether the two crops are different selective environments for pea aphids, we can compare clonal rates of increase for replicates of the same clones reared on alfalfa and clover.

Despite the fact that genetic variability in life tables is the basis for life-history evolution, there are surprisingly few experimental estimates of complete life tables for different genotypes (Charlesworth 1984). In the study described here and in Via (1991a), complete individual life tables for clonal replicates that were each allowed to develop on one of the two host plants were constructed. In addition to revealing the population structure of host adapatation in which we are interested, these genotype-specific life tables provide a rich source of data for the genetic analysis of the evolution of life histories.

The Experimental Organism

The pea aphid (*Acyrthosiphon pisum*) is a pest of leguminous crops. In the dairy-farming areas in which these experiments were performed, alfalfa and red clover are by far the major hosts. Peas are not planted in these areas, and I have not found pea aphids on roadside vetch. In the study sites in Iowa and upstate New York, alfalfa and red clover are planted in a dense patchwork with fields of these hosts interspersed with nonhosts such as corn or woodlots. In Johnson County, Iowa, a pasture mix of alfalfa, red clover, and birdsfoot trefoil was occasionally found (alfalfa, clover, and mixed fields were in about a 6:3:1 ratio). In Tompkins County, New York, no mixed fields were found in the study area until one was located in 1992 (ratio of alfalfa-clover acreage is about 5:1). Because well-defined patches (fields) of these two crops occur together in local patchworks, this is a good model system for the study of the evolution of population divergence and local adaptation.

Pea aphids are cyclically parthenogenetic. In areas with cold winters such as Iowa and New York, declining photoperiods and temperatures in the fall stimulate the production of sexual forms (Smith and MacKay 1990). The males and females of a given clone are genetically identical to one another, and recombination in the gametes breaks up the clonal genotype and generates genetic variability among the overwintering eggs. Because clones are not necessarily homozygous, recombination can generate variable progeny even when males and females of the same clone mate (Via, unpublished data). The eggs hatch in the spring into "fundatrices," which establish clonal lineages. In the fundatrix generation, clonal diversity is expected to be at its peak because of the genetic variation released by recombination the previous fall (Lynch and Gabriel 1983). No recombination occurs in the clonal phase (Blackman 1979). If recombination did sometimes occur, we would expect to see segregation at heterozygous allozyme loci. However, repeated electrophoretic analysis of the same clones in laboratory culture over a two-year period has produced no hint of such segregation (Via, unpublished data).

A cyclically parthenogenetic crop pest like the pea aphid has some features that facilitate the study of population structure and local adaptation:

1. Because of the clonal mode of reproduction during the summer months, it is easy to test exact genotypic replicates in each environment. This simplifies the quantitative genetic analysis used to estimate genetic variation because no mating design is necessary (for a similar analysis on a sexual species, see Via 1984a,b, 1988; Groeters 1988). Note that the clonal component of variance from a genetic analysis is the total genetic variance and includes both additive and nonadditive genetic variability. However, because clones are completely linked groups, the total genetic variance is what determines the response to clonal selection during a single summer season.

2. Clones can be cultured in long-day conditions in the laboratory and prevented from entering the sexual phase. Thus genotypes can be kept intact and we can go to our cultures and retest interesting clones or use the same genotypes for other experiments. This would not be possible in a sexual species in which genotypes are broken up every generation by recombination.

3. We can induce the sexual forms in the laboratory. This allows us to make crosses to investigate the underlying genetic basis of host specialization and other characters of ecological importance.

4. Pea aphids are found in a simple agricultural system, with two host plants, one main fungal pathogen, two main parasitoids, and several generalist

predators. In such a simple system, it is possible to study the interspecific inter-actions that shape the evolution of the species. We know so little about the evolutionary dynamics of interactions among species that it is quite an advan-tage to be able to study them in a simple system before we attempt to under-stand some of the very complicated interactions in natural communities. Al-though the food resource in agricultural systems is clearly manipulated, the dynamics of pea aphid populations and those of their parasitoids and patho-gens are generally unmanaged. *A. pisum* is not a major pest, and so insecticides are rarely applied for its control. Because alfalfa and clover are low-value crops, insecticide use is, in general, fairly low (most farmers in our area do not spray at all). This means that the population dynamics of pea aphids are not driven by insecticide treatments: they are essentially a "natural population." Thus this system is a good model for the study of evolutionary interactions among insects, their host plants, and their natural enemies.

The rest of this chapter consists of a description of four experiments. In experiment 1, I determined the extent of genetic structure in pea aphid populations on two hosts using a reciprocal transplant under field condi-tions. Experiment 2 tested the hypothesis that the host plant specializa-tion observed in the reciprocal transplant might be due to experience rather than genetics. Experiment 3 concerned an analysis of sexual crosses among a set of clones that spanned the range of performance on the two crops, and experiment 4 was a test of the relative demographic performance on two crops of progeny produced by crossing an alfalfa specialist with a clover specialist.

Experiment 1: Analysis of Population Structure

Experimental Design

To estimate the genetic structure of host use in an area in which both alfalfa and red clover are grown, a reciprocal transplant was performed in which the relative performance of clones collected from alfalfa and clover fields was measured on each of the two crops (fig. 4.1). For this experiment, a hierarchical sampling design was employed in which multi-ple clones were collected from each of several fields of the two hosts within a very localized geographical area. This design allows genetic vari-ation to be partitioned into components that represent (1) variation be-tween groups of clones collected from different source crops, (2) variation between groups of clones collected from different fields within source crops, and (3) variation among clones within fields.

To test the performance of each clone on each of the two crops, a reciprocal transplant design was used, with the modification that the

FIELD EXPERIMENT:
EXPERIMENTAL DESIGN

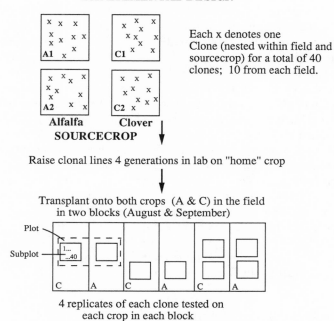

Each x denotes one
Clone (nested within field and
sourcecrop) for a total of 40
clones; 10 from each field.

Alfalfa   Clover
SOURCECROP

Raise clonal lines 4 generations in lab on "home" crop

Transplant onto both crops (A & C) in the field
in two blocks (August & September)

4 replicates of each clone tested on
each crop in each block

FIG. 4.1. Schematic of the collection and testing design. Because four sets of plots were needed, but only three paired strips of alfalfa and clover were available, two sets of subplots were located within one of the paired strips. See text for further explanation. (Reprinted with permission from Via 1991a)

clones sampled on different farms were tested in a single test field in which alternating strips of the two crops had been planted. This permitted testing of clones from all collection sites in a single "common garden" that contained both crops. Relative to a transplant between the original fields, the use of a single site containing both crops served to control environmental variability not related to the crop and to minimize travel time.

Sampling Clones from Field Populations

In this experiment, seven to ten parthenogenetic pea aphids were collected from each of two alfalfa fields and two clover fields that were located in a small (<10 square miles) area in Johnson County, Iowa (fig. 4.1). Collections were made relatively early in the season (late May), about two generations after egg hatch, which would probably have occurred by early April. Thus variability among clones would still be expected to be high, but not as high as if collections had been made during the first clonal generation before any selection had occurred. For this reason, estimates of variation among clones from this experiment probably underestimate the genetic variation present at the beginning of the season. In all collections, efforts were made to collect widely throughout a given field in order to maximize the opportunity of collecting different clonal genotypes if they were present.

The aphids were returned to the laboratory and were placed individually in vented 5L plastic buckets that contained three potted alfalfa or clover plants each. All aphids were housed on the same crop from which they had been collected (henceforth, the "home plant"). Housed in this way, each field-collected aphid established an independent clonal lineage. The possibility that maintaining aphids only on the home plant might cause apparent specialization due to experience in the absence of genetic differences between clones was tested in experiment 2. Lineages were subcultured and transferred to new containers with fresh plants every fifteen days. Rearing conditions were 20°C and sixteen hours of light/day in a controlled-temperature plant-growth chamber (EGC, Chagrin Falls, Ohio). Aphid clones were maintained in this way for approximately four generations, during which time no clones were lost.

Field Test of Relative Clonal Demography on Two Hosts

This experiment entailed evaluation of the relative lifetime fitness of each clone from the sampling design on each of the two crops under field conditions. This provides the picture of population structure in which we are interested. To this end, a two-acre test field was planted into six alternating strips of alfalfa (var. "Duke"), and red clover ("Medium Red"). For each of two time blocks (the experiment was repeated in August and September), four paired subplots of alfalfa and clover were established (fig. 4.1). In each subplot, one small organdy bag was placed on each of forty evenly spaced plants about one week after the field had been cut, when the plants were about five inches high. These bags were supported by a plastic stake and were tightly sealed on the top and bottom around a foam plug to prevent the escape of the test aphid and its progeny. Each bag contained several sprigs of foliage, including leaves, stems, and buds.

Clones were kept synchronized in the lab so that adults of every clone would be available on the same day for transport to the field. One copy of each clone was tested in each subplot (fig. 4.1), for a total of 4 clonal replicates on each crop in each of the two time blocks. Clones were assigned to plants at random within each subplot. For each test, one adult female of a specified clone was placed in each bag. The next day, bags were checked, and if the adult had produced offspring, she and excess offspring were removed, leaving two offspring in each enclosure. After five days, one of these was removed according to an arbitrary rule, leaving a single test individual in each enclosure.

Complete life tables were made for each test individual. Age-specific survival was obtained by checking the enclosures every two days to see whether each individual was still alive. Age-specific fecundity was measured by transferring the test individuals to new enclosures on a nearby plant every four days after they became adults. The old enclosures were then collected and offspring were counted in the laboratory. In all, 640 individual life tables were made in this experiment.

Demographic Analysis of Fitness

From the data required to construct the life table for each individual, we also obtained measures of life-history components such as longevity and total fecundity. To obtain a measure of individual fitness in a demographic context for the kth replicate of the jth clone on the ith crop, I iteratively solved the Euler equation. This provides an estimate of λ_{ijk}, which can be interpreted as the rate of increase that an ideal population comprised only of the demographic phenotype of the kth individual of the jth clone on the ith crop would attain at stable age distribution:

$$1 = \sum_{x=0} l_{ijk}(x) f_{ijk}(x) \lambda_{ijk}^{-(x+1)}. \tag{4.1}$$

In eq. (4.1), $l_{ijk}(x)$ is the probability that the test individual survived to day x (for an individual this is either 1 or 0), and $f_{ijk}(x)$ is the number of offspring produced by the test individual in the xth interval that survived to be in the 0th age class, at which time the test individual was in age class $(x + 1)$ (Michod and Anderson 1980).

Theory developed by Charlesworth (1980) reveals the consequences of clonal variation in the rate of population increase. However, the best way to estimate the clonal rate of increase from experimental data is not clear. From the individual life tables, estimating the clonal rate of increase is complicated by environmentally caused variability among the life tables of the genetically identical clonal replicates. When some, but not all, replicates of a given clone die before reproduction (perhaps due to being placed in an enclosure in a poor microsite), the rates of increase calculated by different methods can differ rather considerably. Picking the most suitable method requires knowledge of how individual aphids might experience environmental variation in a field situation (Via, in prep.). There are three possible methods for estimating the clonal rate of increase: (1) estimate the clonal rate of increase ($\lambda_{ij.}$) from the mean life table for each clone constructed by averaging the individual $l_{ijk}(x) f_{ijk}(x)$ values for each

x over all replicates of the clone; this is appropriate if one believes that individuals move frequently among plants, thus truly averaging over plant-to-plant variation; (2) take the maximum rate of increase attained by any of the clonal replicates, $\lambda_{ijk(max)}$, under the assumption that individuals that are free to move between plants will eventually find a plant that is at least as good as the best one(s) experienced in this experiment; (3) take the mean of the individual rates of increase; this is most appropriate if individuals do not choose the plant on which they feed. Method 3 is useful for the study of variability because the individual variation among replicates is maintained in the data set, so that variation in the rate of increase among clones can be partitioned into genetic and environmental components using standard ANOVA techniques in the same way as one would partition variation in the life-history components. However, this method is probably not as accurate for projections as the other methods because it tends to overestimate the effects of replicates that die before reproduction. This can cause some clones to have values of λ_{ij}. that are <1 even though some replicates reproduced successfully.

Statistical Analysis of Population Structure for Quantitative Traits in Different Environments

A complete picture of the population structure of demographic performance in two environments can be obtained from a hierarchical set of analyses on the data from the reciprocal transplant. This set consists of analyses of variance (ANOVAs) performed on the following data sets: (1) on the full data set of all clones tested on both crops (the "full" model); (2) on all clones tested on each crop individually (the two "Testcrop" models); and (3) on subsets of clones from each of the two crops when tested separately on the two crops (the four "Sourcecrop * Testcrop" models). A summary of the hypotheses tested in each of these analyses can be seen in table 4.1.

Note that in the "full" model (table 4.1A), the interpretation of the main effects can be complex. For example, Fry (1992) points out that in a simple two-way ANOVA (e.g., with factors "Family," "Testcrop," and "Family × Testcrop"), the "Family" main effect can test two different hypotheses, depending on which factor is used in the denominator of the F-test. If the "Error" mean square is used, the "Family" main effect is interpreted as the variance among families in the frequency-weighted "average" of the environments tested (which, as Fry points out, is unlikely to provide insight into genetic variation under field conditions where the frequency and range of environments experienced are almost certainly different from those in the lab). If, however, the "Family ×

TABLE 4.1
Questions addressed by each term in the different analyses of variance used to evaluate population structure in the pea aphid

Source	Meaning of F-Test
A. "Full Model"	
Sourcecrop	Do collections from alfalfa and clover differ in performance averaged over both test crops?
Field(Sourcecrop)	Do collections from different fields within crops differ in performance when averaged over test crops?
Clone(Field, Sourcecrop)	Do aphid clones within fields and source crops differ in performance on the "average" crop?
Testcrop	Are alfalfa and clover different environments for the "average" pea aphid?
Sourcecrop × Testcrop	Do aphids from the different source crops have different relative performance on the two test crops? This is the test for local adaptation.
Field(Sourcecr) × Testcrop	Do aphids from different fields within crops differ in relative performance on the two hosts?
Clone(Field, Source) × Testcrop	Do aphid clones within fields and source crops differ in relative performance on the two hosts?
Error	Variation among replicates within clones.
B. "Testcrop" Models (one for each test crop)	
Sourcecrop	Do aphid clones from alfalfa and clover differ in performance on one of the test crops?
Field(Sourcecrop)	Do aphid clones from different fields within source crops differ in average perfromance on the test crop under consideration?
Clone(Field, Sourcecrop)	Do aphid clones within fields differ in performance on one of the test crops?
Error	Variation among replicates within clones.
C. "Sourcecrop × Testcrop" Models (one for each of the four combinations of collection crop and test crop)	
Field	Do aphid clones from the two fields of the source crop in question differ in performance on the test crop under consideration?
Clone(Field)	Do aphid clones within fields of the source crop in question differ in performance on the test crop under consideration?
Error	Variation among replicates within clones.

Notes: ( ) denotes "nested within". Note that the main effect in the "Full" model (A) can also be interpreted in terms of covariances across environments, depending on the construction of the F-tests (see text and Fry 1992).

Testcrop" interaction is used as the denominator of the F-test, the "Family" term tests for a non-zero covariance of family values across environments. Presumably the same two classes of hypotheses could be tested for the main effects in the more complicated nested and crossed model presented in table 4.1A.

The hierarchical approach to the analysis of a reciprocal transplant that is presented in table 4.1 is useful because no single analysis permits all of the interesting hypotheses to be tested. The full model (table 4.1A) allows the *relative* performance of aphids from different clones, fields, and source crops to be evaluated on the two crops (using the interaction terms). I do not generally use the main effects in the "full" model. Instead, estimates of variation among aphid clones in performance are obtained from analyses of observations on each crop taken separately (table 4.1B). Finally, the analyses by source crop and test crop (table 4.1C) permit the variation among aphids from each source group to be analyzed separately on the "home" and "away" hosts.

Results

POPULATION STRUCTURE

At the population level, the mean longevity, fecundity, and rate of increase for the clones from each collection field on these two crops revealed a striking degree of local adaptation (fig. 4.2). On average, clones showed much better performance on the "home" crop than they did when tested on the "away" crop. This resulted in a highly significant "Sourcecrop × Testcrop" interaction for all the demographic variables (see Via 1991a for complete ANOVA tables). In general, the interaction between collection environment and test environment is the term used to test for local adaptation in quantitative characters (to be sure that the interaction is in the appropriate direction, it is advisable to plot the data as in fig. 4.2).

Despite the highly significant degree of local adaptation, there was also significant variability in the relative performance of clones within fields on the two test crops. This can be seen in the scatterplots of clone means shown in figure 4.3. These scatterplots also clearly show the divergence between the groups of clones collected from the two crops. Variation in the relative performance of clones within fields on the two test crops was estimated using the "Clone(Field,Sourcecrop) × Testcrop" interaction in the analysis of variance for the full model. This interaction revealed significant variation among clones within fields in relative age at first reproduction, longevity, and rate of increase on the two crops (see Via 1991a).

Analyses of variance on each test crop separately revealed that there were highly significant differences in the average performance of clones

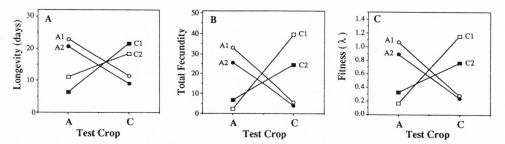

FIG. 4.2. Mean longevity, total fecundity, and fitness (λ) for each collection field when tested on alfalfa and clover. Points were calculated as the mean over all clones within a field. Circles represent means of subpopulations collected from the two alfalfa fields; squares are means of subpopulations from clover. For all variables on both test crops, means of aphids collected on alfalfa differ from those of clones originating from clover (see Via 1991a for tests). (Reprinted with permission from Via 1991a)

collected from the two source crops on both alfalfa and clover (see Via 1991a for full details of the analysis). This was expected, given the local adaptation seen in the full analysis and in the plotted data (figs. 4.2, 4.3). There was also variability among clones from the same source crop for most of the demographic characters on both alfalfa and clover. This within-crop genetic variation is the raw material for the evolution of even greater specialization in this system. Variability in performance of clones on the individual test crops can be visualized in the graphs by projecting the bivariate scatter in fig. 4.2 onto each axis.

Note that in a hierarchical sampling design such as this, in which clones are nested in fields and fields are nested in source crops, variation in performance is partitioned into components among clones within fields, between fields within crops, and between subpopulations collected from different source crops. This is all genetic variation. Purely environ-

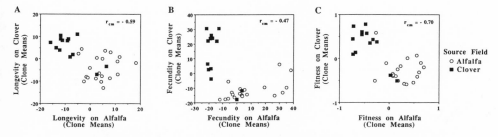

FIG. 4.3. Scatterplots of clone mean performance on alfalfa and red clover when each clone was tested on each of the crops. Clone means were obtained by averaging individual values over 4–8 replicates. Points are residuals from model that removed the effect of time block. r_{cm} is the cross-crop correlation of clone means. These distributions show the range of clonal variability in the "regional" population. (Reprinted with permission from Via 1991a)

mental variability is found in the variance of clonal replicates, that is, in the within-clone term, or the bottom-most mean square in the analyses of variance.

TRADE-OFFS IN PERFORMANCE ON THE TWO CROPS

The criterion for local adaptation is that subpopulations have greater performance on the crop from which they were collected than do clones collected from the other crop. The clones that were tested here clearly satisfied this criterion, exhibiting strong local adaptation (figs. 4.2, 4.3). This pattern of locally adapted performance produced a pronounced negative clonal correlation in fitness (λ) on the two crops in this collection of clones from two hosts. The negative genetic correlation in λ across crops seen in the regional collection (correlation of clone means with all clones included, $r_{cm} = -0.70$ [−0.84, −0.47]; Via 1991a) has clearly been enhanced by the divergence between the means of the subpopulations on the two hosts. However, it is unlikely that divergent selection in the two crops would have led to such a pronounced negative genetic correlation unless the within-subpopulation genetic correlation, on which selection acts, were at least slightly negative. Estimation of the average cross-crop correlation in fitness ($\lambda_{ij.}$) within subpopulations revealed that it was negative, but of much lower magnitude than the correlation seen at the regional level ($r_{cm} = -0.38$ [−0.64, −0.05]; Via 1991a). Thus, there may be some fundamental genetic constraint, such as covariance of the effects of new mutations (cf. Lynch 1985) that limits the possibility of specialized performance on two hosts. Generalists with high performance on both hosts were simply not found in field collections. However, as we will see in experiment 4, clones with generalized demography can be generated if alfalfa and clover specialists cross.

Experiment 2: Can Host Specialization Be Altered by Experience?

Experimental Design

Results of the field study discussed above showed that clones tended to be highly specialized and to perform better on the "home" than on the "away" host. However, because the clones used in the field experiment were all housed in the laboratory on the "home" host, there was some possibility that the observed differences might have been due to experience rather than to genetic differences among the clones. The purpose of experiment 2 was to test this hypothesis.

EXPERIMENTAL DESIGN

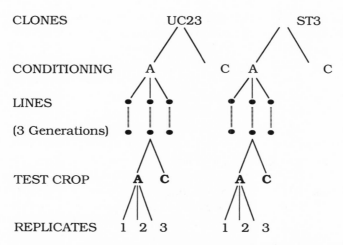

Fig. 4.4. Experimental design for experiment 2, which tested effects of experience on host specialization. For each of two clones, three conditioning lines were maintained on each crop for three generations. A = alfalfa, C = clover. UC23 was a clover specialist, ST3 was an alfalfa specialist. Three replicates from each conditioning line were tested on each crop. Longevity and total fecundity were measured for each replicate. (Reprinted with permission from Via 1991b)

Two clones that showed specialized host performance were chosen from the group tested in experiment 1. Because this design (fig. 4.4) entailed rearing each of the clones on both hosts, the most extreme clones were not chosen because they generally could not be maintained on the "away" crop. However, the slightly less specialized clones used here are most likely to have been the ones in the field that would have been influenced by experience, because they could at least persist for several generations if they were to migrate to the other crop. Thus, they provide the information that we seek on the possible modification of host specialization by experience.

Three lines of each of these clones were established on each of the two crops so that both clones would have some replicates that had experienced the same host (fig. 4.4). These "conditioning" lines were reared for three generations. This time period was chosen because three generations appears to be the span over which maternal effects may extend (Lynch 1985). After this period, replicate individuals from each of the conditioning lines were tested on each of the two crops in a complete factorial design.

Complete life tables were constructed for individuals as in experiment 1, except that aphids were raised in the laboratory on greenhouse-grown plants rather than in the field. By monitoring each individual for its entire lifetime, we obtained replicate measures of longevity and lifetime fecundity for clones from each conditioning treatment when tested on each of the two crops. Data were analyzed with a three-way fixed-effects ANOVA, in which the factors were "Clone," "Conditioning crop," and "Test-crop."

Results

There was no evidence of a significant effect of the conditioning crop on the relative total fecundity of clones on the two test crops ("Conditioning crop × Testcrop," $p > 0.35$; "Clone × Conditioning crop × Testcrop," $p > 0.5$). In keeping with the expectation from the field results (fig. 4.5I), the two specialized clones maintained their relative fecundity on alfalfa

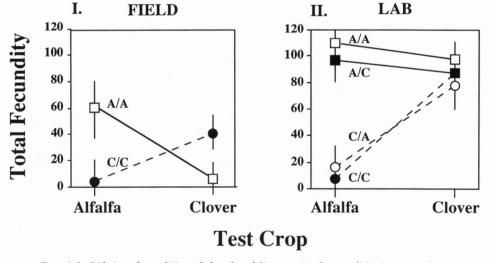

FIG. 4.5. Lifetime fecundities of the clonal lineages in the conditioning experiment when tested on alfalfa and clover. The clone with specialized performance on alfalfa is denoted by square symbols and solid lines; the clone that is specialized on clover is marked with round symbols and dashed lines. The first letter of each two-letter code indicates the crop on which the clone was specialized, and the second letter indicates the crop on which the clone was maintained. Thus, A/A and C/C mark lines that indicate the fecundities of clones when maintained on the "home" crop. A/C and C/A mark the fecundities of clones when raised on the "away" host. I. Results for these two clones from the field experiment described in experiment 1. II. Total fecundities on each crop for the different conditioning lines in experiment 2. (Reprinted with permission from Via 1991b)

for both conditioning treatments (fig. 4.5II). Unfortunately, however, we were unable to evaluate the effects of conditioning on performance on clover, because both clones had uniformly high fecundity on the greenhouse-grown clover plants. Nevertheless, the clonal rankings on alfalfa provide evidence that experience alone is not enough to produce specialized performance of the magnitude observed in the field in experiment 1 (see also Via 1991b).

Experiment 3: The Genetic Architecture of Specialization

Results of the field experiment to evaluate population structure revealed that the collections from alfalfa and clover formed essentially two distinct performance groups: one group that was specialized on alfalfa, and another that was specialized on clover. Between these two groups is a gap in which few generalists with intermediate performance were observed (see fig. 4.3C). Why was the distribution of relative performances on the two crops so discontinuous, and why were no intermediates observed?

There are at least three hypotheses for the absence of intermediate forms: (1) The groups on alfalfa and clover are reproductively isolated, so crosses between extremes never occur and intermediates are not generated; (2) intermediates are never generated, even if crosses of extremes occur, because of the genetic architecture of host plant specialization; (3) divergent selection in monocultures of alfalfa and clover eliminates the intermediates because they do not have as high a rate of increase on either crop as do the more specialized clones.

Experiments 3 and 4 were designed to test these hypotheses. These experiments involved making a series of crosses and then testing the demographic phenotypes of the progeny clones on the two crops.

Experimental Design

Six clones that had been collected in Tompkins County, New York, in 1987 and field-tested for demographic performance on alfalfa and clover were used in this experiment. They were chosen to span the same range of demographic variability seen in figure 4.3. They included two clones that were highly specialized on alfalfa, two clones that were highly specialized on clover, and two clones that were as intermediate as possible. These clones were induced to produce sexual forms by manipulating photoperiod and temperature (Via 1992). Males and females of all clones were produced and were available contemporaneously so that all possible combinations of matings could be made. For each cross between a given pair of clones, several replicates were made that involved different (genetically identical) individuals from each clone. Reciprocals were also made for each cross.

Each replicate cross consisted of confining one male with three females of the designated clones in a small dish containing sprigs of the aphids' home plant(s). Eggs were removed from the dishes weekly, and were surface-sterilized to reduce the possibility of fungal contamination during the long cold period that was required to break diapause and obtain reliable egg hatch (see Via 1992 for details of this procedure). After the required cold treatment, eggs were brought back to room temperature and allowed to hatch.

Results

On average, the percentage of the eggs that hatched was very high (60%), and in some individual crosses over 89% hatched (Via 1992). Hatching success varied widely between crosses. In particular, in selfed crosses between males and females of the same clone, only about half as many eggs hatched as in crosses between clones that were different genotypes but were in the same performance class (Via, in prep.). This indicates the presence of inbreeding depression in these aphid populations, and suggests that intraclone mating may not be the rule in field situations.

Results of these crosses also showed that crosses between clover and alfalfa specialist clones tended to produce somewhat fewer eggs than did crosses between clones of the same performance phenotype (mean of 6.3 eggs/female compared with 12.2 eggs/female for the same-phenotype crosses; Via, unpublished data). The eggs from these crosses also had a lower overall success rate in survival from egg to adult than did crosses within specialist groups (Via, in prep.). These results suggest that there may be partial pre- and postmating reproductive isolation between the subpopulations of pea aphids on alfalfa and clover. However, it is clear that clones from alfalfa and clover are still capable of mating and producing viable progeny. The frequency with which this occurs in the field is currently unknown.

Experiment 4: Demographic Variation among Progeny from Crosses of Specialist Clones

Experimental Design

The crosses described in experiment 3 included several crosses between clones that were extreme specialists on different crops. From one such cross, twenty-eight progeny aphids were kept (nine had the clover clone as the mother, and nineteen were from the reciprocals in which the alfalfa clone was the mother). From these progeny, clonal lineages were established so that five replicates of each of the progeny genotypes could be tested on each host. All progeny clones plus the two parental clones were maintained for three generations prior to the experiment in 5L buckets that contained one plant of each host. For each clonal replicate, longevity and total fecundity were measured for individuals as in experiment 2. As controls, five replicates of the parental clones were also tested on each host.

Results

Analysis of the demographic performance of the progeny clones revealed that intermediates are certainly generated by recombination and segregation in crosses between clones with extremely specialized host performance. The mean longevity and fecundity of the progeny on each host as compared to the parental values is shown in table 4.2. For all of the variables except longevity on clover, progeny means were statistically indistinguishable from the mean of the parental values.

There was significant variation among the progeny clones for longevity on alfalfa and fecundity on both crops (Via, in prep.). This clonal genetic variation in the F_1 shows that if the same loci influence adaptation to alfalfa and clover, the parents were not homozygous for alternative alleles because the progeny of homozygous parents would all be heterozygotes, among which there would be no genetic variation. Alternatively, different loci may influence performance on the two crops, in which case heterozygote progeny would be generated even if parents were fixed at all relevant loci.

The results of these crosses suggest that successful matings can occur between specialists and that there is no genetic barrier to the generation of genotypes with intermediate performance on the two crops. Thus the

TABLE 4.2

Comparisons of performance of specialist parents and their offspring on two hosts, alfalfa (A) and clover (C)

Group	Longevity (A) (days)	Longevity (C) (days)	Total Fecundity (A)	Total Fecundity (C)
Alfalfa specialist	37.6	9.1	101.8	4.0
Clover specialist	9.3	33.3	1.3	80.5
Parental mean	23.5	21.2	51.6	42.2
Offspring mean	25.8	30.9*	52.4	44.3

* Differs from parental mean at $p < .01$.

absence of such intermediates in field samples must be due either to an infrequency of matings between specialists or to selection against inter-mediate forms because they are not competitive with specialists on either crop.

Discussion

The study of patterns of genetic variation within species can provide fun-damental insights into the process of evolution. When a species inhabits several different environments in a localized area, genetic divergence among subpopulations in relative performance in the different environ-mental types can reveal local adaptation. The discovery of local adapta-tion in these closely adjacent subpopulations of pea aphids not only shows that the habitats are distinctly different selective environments, but suggests that genetic variation for performance in the different environ-ments has been available in the past to produce a localized response to selection of a magnitude sufficient to overcome whatever gene flow may occur.

In the experiments described here, a hierarchical sampling design was used to partition genetic variation in host-plant exploitation within and between subpopulations of pea aphids collected from two host crops. Testing replicates of the same aphid clones on both crops in a reciprocal transplant design permitted the estimation of the relative performance of genotypes in two environments and was an essential manipulation in the investigation of local adaptation. This analysis of the genetic structure of pea aphid populations revealed that subpopulations from two closely ad-jacent hosts are genetically divergent: aphid clones are highly locally adapted to the crop from which they were collected.

In addition, by illustrating that different clones had high fitness on each of the two hosts, the reciprocal transplant experiment showed clearly that alfalfa and red clover are different selective environments for pea aphids. We currently do not know what the difference is between the clones that do well on alfalfa and those that do well on clover. Perhaps clones differ in the morphological traits necessary to navigate on or cling to plants that differ in the density and length of trichomes (alfalfa has few short tri-chomes and clover has many long trichomes). Or, the plants may differ in chemicals which are attractive or repellant, with corresponding differ-ences in the sensory systems and/or behavior of the aphids that success-fully exploit them. An understanding of the morphological, physiologi-cal, and behavioral mechanisms involved in host specialization would be very interesting, and would reveal the phenotypic characters that are the targets of natural selection on these two hosts.

The results of the conditioning experiment (experiment 2) support the conclusion that the differences between clones in performance on the two hosts seen in the field experiment are genetic and not caused by experience. Even three generations of experience on the "away" crop were not sufficient to reverse the specialization seen in the field experiment. One might argue that the results from experiment 2 (fig. 4.5) show a trend toward modification of specialized performance by experience, suggesting that perhaps a larger effect might be seen after additional time in the conditioning treatments. However, it is unlikely that a clone with extremely poor initial performance on the "away" crop would persist more than three generations there, so it is questionable whether continuing the treatments for a longer period would be biologically meaningful. Moreover, the differences in performance on the two hosts between the specialist parents and their offspring, which were all cultured in a common environment (experiment 4), are further evidence that the host specialization that we observed is due to genetic differences between clones rather than to environmental influences.

From the inbreeding depression seen in the selfed crosses in experiment 3, we can infer that selfing may not be common in field populations. Further evidence that inbreeding is uncommon was provided by the crossing studies, which showed that the F_1 progeny of crosses between specialists on different hosts are genetically heterogeneous and thus suggested that the most specialized genotypes currently found in these aphid populations are not homozygous. Therefore, even though inbreeding might hasten the divergence of populations by increasing the homozygosity of alleles involved in host specialization and limiting gene flow, it is evidently not occurring with great frequency in this pea aphid population.

Open Questions and Future Directions

The very poor performance of individuals transplanted to the "away" crop suggests that strong disruptive selection is acting in these two crops. In addition, a small but significant level of reproductive isolation was seen between clones from the two hosts. However, these divergent populations are by no means genetically homogeneous. Considerable genetic variation was found among clones sampled from the same host, and the crosses revealed that even the most highly specialized genotypes may be heterozygous at a number of loci influencing host performance. How is the genetic variability that I observed maintained in the face of strong selection within crops? Are the subpopulations on alfalfa and clover in the process of speciating?

Because gene flow between these two crops would both maintain variation and constrain speciation, determining the extent of migration and gene flow between crops will be essential for answering these questions. It is also important to determine whether alate migrants can choose their habitat, because habitat choice would reduce gene flow and thus facilitate sympatric speciation (Maynard Smith 1966; Felsenstein 1981; Rice 1987).

It is important to note that characters other than host use are probably also under selection in pea aphid populations. For example, heat and desiccation (which are hazards when the hay is cut) or natural enemies might act as additional selective forces. If the susceptibilities to these selective agents are genetically correlated with host use, then selection on them may affect patterns of clonal variation for host use. Could genetic correlations among characters relevant to the different forces of selection constrain the overall response to multivariate selection and slow the loss of variance? Might natural enemies act in a frequency-dependent way and thus preserve genetic variability? These are ecological as well as genetic questions.

I will now consider several of the factors potentially involved in the maintenance of clonal variation in more detail.

MIGRATION AND GENE FLOW

The observation that a few of the clones that were collected for experiment 1 actually performed better on the alternate crop (see fig. 4.3) suggests that they may have been either very early migrants or progeny of individuals that migrated the previous fall. Other than this suggestion that migration does occur, virtually nothing is known about the extent of gene flow between the subpopulations of pea aphids on different crops.

Results of the reciprocal transplant clearly suggest that migration between crops would have deleterious selective effects on the individual fitness of migrants. Because most clones had lower fitness on the "away" host than did the resident clones, migrants to a field of the other crop would be at a disadvantage relative to the residents. In addition, individuals placed on the "away" host often become shriveled and appear weakened (Via, personal observation), which might make them more susceptible to natural enemies than they would be on the "home" host.

Because of these disadvantages to migrants, migration at different times of the season may have very different effects on gene flow. We know that alate formation has two seasonal peaks, one in June and the other in early September, and trapping studies suggest that movement is greatest at these two times also (Via, unpublished data). Migrants during the first peak of movement must survive seven to eight subsequent clonal generations before sexuals are produced. Thus, it is unlikely that early-season migrants between crops, especially those that were most specialized on the other crop, would contribute much to the mating pool that is formed during September. In contrast, late-season migrants, particularly mothers of the sexual forms (which may be winged) or the sexual males themselves, have a much higher probability of surviving long enough to mate and bring about gene flow before being eliminated from the population. The sexual females are all wingless, so they are not capable of migration. Although these late-season migrants may be quite specialized in their host-plant performance, and thus may have low survivorship on the alternate host, any matings that do result will be quite significant in terms of gene flow.

The frequency of matings between clones that are as extreme in their level of host specialization as those used in the crosses (experiments 3 and 4) is unknown. However, when such matings do occur, our tests of the progeny of extreme parents showed that clones with intermediate performance on both hosts will be generated. The fact that we have not seen many intermediate clones in our collections suggests that, if produced by cross-crop matings, they may be eliminated by selection in the monocultures usually planted by farmers because their fecundity is not as high on either host as is that of the more specialized resident. This hypothesis could be tested with a selection experiment in which a mixture of specialized and intermediate clones was placed in a population cage in which population size could be kept in check by random mortality introduced by the experimenter. Would the specialist clones really come to dominate the population in a monoculture? It would also be interesting to sample from fields in which a mixture of alfalfa and clover had been planted. Are there more intermediate clones in such fields, or are they simply mixtures of specialists that choose their hosts?

SENSITIVITY TO HEAT AND DESICCATION

Farmers try to cut alfalfa when conditions are hot and dry, so that the hay can be rapidly baled and removed from the fields. After cutting, the amount of edible foliage and shade is dramatically reduced and the aphids that survive are exposed to both heat and desiccation stress. Regular sampling during the 1988 season revealed that aphid densities drop drastically after cutting (Hural and Carruthers, in prep.). Are the aphids that survive really a random sample of those present before cutting, or is there differential sensitivity to heat and the starvation that may occur during the five to seven days before the alfalfa plants sprout again? Preliminary experiments on this issue suggest that clones may differ in sensitivity to starvation and desiccation at high temperatures (Alison Shaw, unpublished data). Thus, cutting the crop may cause additional selection on pea aphid populations.

Herbivores are exposed to more than just their host plants in field situations. Even simple agricultural systems such as this one contain natural enemies. If variation in susceptibility to predators, parasitoids, or disease exists in herbivore populations, then these natural enemies will act as agents of selection. Work in progress indicates that pea aphid populations contain significant clonal variability both for susceptibility to a wasp parasitoid, *Aphidius ervi* (Henter and Via, in prep.), and to a fungal pathogen, *Pandora neoaphidus* (Hural, Carruthers, and Via, in prep.). Preliminary studies of the aphid-parasitoid-disease interaction also suggest that there is a weak negative genetic correlation between susceptibility to the parasitoid and to the fungus (Hural, Henter, and Via, in prep.). Such a correlation would, of course, mean that simultaneous response to selection by this parasitoid and fungus would be slowed (though not entirely prevented; e.g., Slatkin and Kirkpatrick 1987).

Conclusions

Clearly, understanding the dynamics of evolution in even a simple system such as this one requires that one know a great deal not only about the genetics but also about the ecology of the organisms in question. This dual nature of the study of the evolution of ecologically important characters is at the heart of ecological genetics (Via 1990). The studies of population structure and the genetic architecture of specialization that are described here only begin to provide a picture of the evolutionary forces at work in this system. Only when we can identify the full range of selective agents acting on this herbivore, measure the genetic variances and covariances for the relevant characters, and integrate these components into testable predictions of evolutionary change will we be acceptably close to understanding the evolutionary dynamics of this one species in its simple community. These predictions will ultimately require field testing because laboratory conditions cannot adequately approximate the range of environmental variability present in the field. Higher environmental variation can dramatically affect the response to selection by lowering heritabilities in the field relative to those estimated under controlled conditions (Falconer 1989).

The patterns of genetic variation and natural selection within species can be used to reveal the evolutionary processes involved in adaptation to differing environmental circumstances. The present diversity of species on earth may certainly have been influenced by macroevolutionary events such as mass extinctions that may act essentially without regard to variability or level of adaptation. However, adaptive evolution does not result from random forces. Adaptation, considered as the "matching of organisms to their environment," or more technically as the evolution of the phenotype toward some optimum value, is the result of natural selection. If studies such as the one described here reveal the manner in which natural selection can create divergence among populations, then surely we will have obtained some insight into how adaptive diversity is generated and maintained.

Acknowledgments

I am very grateful to the people who have helped with these very labor-intensive experiments: many undergraduates both at the University of Iowa and at Cornell University have learned about experimental ecological genetics by working on these projects. Special thanks goes to Jordan West, Jill Rogers, and Erica Street. Kirsten Hural and Heather Henter introduced me to the fascinating problems of natural enemies. Alison Shaw did the starvation/desiccation work. All of the above plus David Hawthorne have provided invaluable assistance in the field and many useful discussions of these ideas. Monica Geber made very useful comments on the manuscript. This research has been supported by a Searle Scholars Award from the Chicago Community Trust, by USDA Competitive Grants 87-CRCR-1-2375 and 88-35713-3747, and by Hatch Project NYC 139419.

References

Blackman, R. 1979. Stability and variation in aphid clonal lineages. *Bio. J. Linnean Soc.* 11: 259–277.

Briggs, D., and S. M. Walters. 1984. *Plant Variation and Evolution.* Cambridge University Press, New York.

Charlesworth, B. 1980. *Evolution in Age-Structured Populations.* Cambridge University Press, New York.

Charlesworth, B. 1984. The evolutionary genetics of life histories. In *Evolutionary Ecology*, ed. B. Shorrocks. Blackwell, London.

Charlesworth, B., M. Slatkin, and R. Lande. 1982. A neo-Darwinian commentary on macroevolution. *Evolution* 36: 474–498.

Conner, J., and S. Via. 1992. Natural selection on body size in *Tribolium*: Possible genetic constraints on adaptive evolution. *Heredity* 69: 73–83.

Falconer, D. S. 1989. *Introduction to Quantitative Genetics*. 3d ed. Longman, New York.

Felsenstein, J. 1981. Skepticism toward Santa Rosalia, or why are there so few species of animals? *Evolution* 35: 124–138.

Fry, J. D. 1992. The mixed-model analysis of variance applied to quantitative genetics: Biological meaning of the parameters. *Evolution* 46: 540–550.

van Groenendael, J., H. de Kroon, and H. Caswell. 1988. Projection matrices in population biology. *Trends in Ecol. and Evol.* 3: 264–269.

Groeters, F. 1988. Relationship between observed components of variance and causal components of variance in a split-family, half-sib, full-sib analysis. *Evolution* 42: 631–633.

Lamb, R. J. and P. J. Pointing. 1972. Sexual morph determination in the aphid, *Acrythosiphon pisum*. *J. Insect Physiol.* 18: 2029–2042.

Lande, R. 1979. Quantitative genetic analysis of multivariate evolution, applied to brain:body size allometry. *Evolution* 33: 402–416.

Lande, R. 1982. A quantitative genetic theory of life history evolution. *Ecology* 63: 607–615.

Lenski, R. E., and P. M. Service. 1982. The statistical analysis of population growth rates calculated from schedules of survivorship and fecundity. *Ecology* 63: 655–662.

Loveless, M. D., and J. L. Hamrick. 1984. Ecological determinants of genetic structure in plant population. *Ann. Rev. Ecol. Syst.* 15: 65–95.

Lynch, M. 1985. Spontaneous mutations for life-history characters in an obligate parthenogen. *Evolution* 39: 804–818.

Lynch, M., and W. Gabriel. 1983. Phenotypic evolution and parthenogenesis. *Amer. Nat.* 122: 745–764.

Maynard Smith, J. 1966. Sympatric speciation. *Amer. Nat.* 100: 637–650.

Michod, R. A., and W. W. Anderson. 1980. On calculating demographic parameters from age frequency data. *Ecology* 61: 265–269.

Rice, W. R. 1987. Speciation via habitat specialization: The evolution of reproductive isolation as a correlated character. *Evol. Ecol.* 1: 301–314.

Schluter, D. 1984. Morphological and phylogenetic relations among the Darwin's finches. *Evolution* 38: 921–930.

Slatkin, M. 1987. Gene flow and the geographic structure of natural populations. *Science* 236: 787–792.

Slatkin, M., and M. Kirkpatrick. 1987. Extrapolating quantitative genetic theory to evolutionary problems. In *Evolutionary Genetics of Invertebrate Behavior*, ed. M. D. Huettel, pp. 283–294. Plenum, New York.

Smith, M.A.H., and P. A. MacKay. 1990. Latitudinal variation in the photoperiodic responses of populations of pea aphid (Homoptera: Aphididae). *Environ. Entomol.* 19: 618–624.

Travis, J., and S. Henrich. 1986. Some problems in estimating the intensity of selection through fertility differences in natural and experimental populations. *Evolution* 40: 786–790.

Turelli, M., J. H. Gillespie, and R. Lande. 1988. Rate tests for selection on quantitative characters during macroevolution and microevolution. *Evolution* 42: 1085–1089.

Via, S. 1984a. The quantitative genetics of polyphagy in an insect herbivore. I. Genotype-environment interaction in larval performance on different host plant species. *Evolution* 38: 881–895.

Via, S. 1984b. The quantitative genetics of polyphagy in an insect herbivore. II. Genetic correlations in larval performance within and across host plants. *Evolution* 38: 896–905.

Via, S. 1988. Reply to Groeters. *Evolution* 42: 634–635.

Via, S. 1990. Ecological genetics and host adaptation in herbivorous insects: The experimental study of evolution in natural and agricultural systems. *Ann. Rev. Entomol.* 35: 421–446.

Via, S. 1991a. The genetic structure of host plant adaptation in a spatial patchwork: Demographic variability among reciprocally transplanted pea aphid clones. *Evolution* 45: 827–852.

Via, S. 1991b. Specialized host plant performance of pea aphid clones is not altered by experience. *Ecology* 72: 1420–1427.

Via, S. 1992. Inducing the sexual forms and hatching the eggs of pea aphids. *Entomol. Exp. and Appl.* 65: 119–127.

Wilson, E. O., and Bossert. W. H. 1971. *A Primer of Population Biology.* Sinauer, Sunderland, Mass.

MICHAEL LYNCH

5

Neutral Models of Phenotypic Evolution

Introduction

For twenty-five years, applications of Kimura's (1968, 1983) neutral model have led to fundamental advances in our understanding of the mechanistic basis of molecular diversity within and between species. Substantial debate still exists over the relative significance of natural selection to patterns of molecular evolution (Gillespie 1991), but the debate itself could never have progressed beyond philosophical discourse had the quantitative predictions of the neutral theory not been available. Regardless of the outcome of this debate, the neutral theory has been established permanently as a null hypothesis for tests of natural selection at the molecular level.

Neutral models have gained prominence only recently in studies of phenotypic evolution. This is surprising since the collection of data at the phenotypic level predates the emergence of molecular biology by decades, and since much of the necessary mathematical machinery of quantitative genetics has long been available. As in the case of molecular evolution, null models for phenotypic evolution can be constructed by considering mutation and random genetic drift to be the only evolutionary forces. Such models provide a basis for evaluating whether observed patterns of variation for quantitative characters are likely to have arisen in the absence of natural selection. In addition, for situations in which effective population sizes are relatively small, so that drift overwhelms selection, neutral phenotypic models can provide first-order approximations for the patterns of genetic variance (Bürger et al. 1989; Houle 1989; Foley 1992). Finally, neutral models are necessary for the analysis of mutation-accumulation experiments used to evaluate the properties of polygenic mutation. Although it is now generally accepted that mutation contributes importantly to the maintenance of genetic variance within populations, there is still considerable disagreement over its quantitative significance (Lande 1975; Turelli 1984; Slatkin 1987a), disagreement that can be resolved only by empirical study.

This chapter provides a brief overview of recent theoretical work on the neutral model of phenotypic evolution. The first part of the paper is concerned with the expected levels of genetic variance for neutral quantitative traits within and between populations. Special attention will be given to the patterns that emerge when sufficient time has elapsed for an equilibrium to arise between the conflicting forces of mutation and random genetic drift. Information on the transient conditions leading up to those patterns can be found in the cited references. Some new results are given for haploids, for asexual species, and for subdivided populations. The second part of the chapter considers how the theory for equilibrium populations can be applied to mutation-accumulation experiments to derive inferences about the properties of polygenic mutation. Several new approaches to parameter estimation are described. In an attempt to make the theory more accessible, many of the technical details leading up to the final results have been omitted.

Expected Patterns of Genetic Variance within and between Populations

Random-mating Diploid Populations

The first case to be considered assumes that mating is random within populations and that gene flow between populations is completely absent. Expanding on earlier work by Clayton and Robertson (1955), Lande (1976), and Chakraborty and Nei (1982), Lynch and Hill (1986) developed a fairly general model for neutral phenotypic evolution for this panmictic type of population structure. They provide results for arbitrary effective population sizes (N_e), and for both monoecious and dioecious mating systems. The results for monoecy adequately approximate those for dioecy provided N_e is greater than a half-dozen or so, which is assumed to be the case in the following.

The model assumes that the population is small enough so that at most two alleles are segregating at any locus within populations. Roughly speaking, this requires that $4N_e\mu \ll 1$, where μ is the genic mutation rate. The number of possible mutations that can arise at a locus is assumed to be effectively infinite. Thus, if allele **A** mutates to allele **a**, and the latter becomes fixed, a subsequent mutation to a third allele can occur, and so on. Significant epistatic effects between mutations are assumed to be absent. The within-locus effects are defined by letting **A** and **a** represent initial and mutant alleles at a locus, and scaling the genotypic effects to be 0 for **AA**, $a(1 + k)$ for **Aa**, and $2a$ for **aa**, with the coefficient of dominance k equaling 0 under additivity, −1 for completely recessive mutations, and +1 for completely dominant mutations.

Regardless of the mating system and population structure, provided the base population is in drift-mutation equilibrium, the population mean genotypic value is expected to change linearly with time under this model. Letting u be the gametic mutation rate (summed over all constituent loci) for the character, for a population with constant size N, $2Nu$ new mutations arise per generation. Each of these has a fixation probability equal to the initial frequency $1/2N$ and, conditional upon fixation, changes the mean genotypic value by $E(2a)$. It follows that each cohort of mutations ultimately causes the expected genotypic value to change by the amount $2Nu \cdot (1/2N) \cdot E(2a) = 2uE(a)$ per generation. In more general terms,

$$\bar{g}(t) = \bar{g}(0) + 2utE(a), \qquad\qquad (5.1)$$

where $\bar{g}(0)$ is the mean genotypic value in the base population, t denotes the generation number, and $E(a)$ is half the average homozygous effect of a mutation. (Throughout this paper, E is used to denote the expected value of a random variable.)

Eq. (5.1) shows that the rate of evolution of the mean phenotype of a neutral character is independent of the population size. Nonadditivity of mutational effects $(k \neq 0)$ does not influence the expected rate of change in the mean between generations either. This is true as long as the expected number of heterozygous loci remains constant, as it will under the assumption of constant μ and N_e, and as long as the spectrum of mutational effects remains constant.

Mutation introduces genetic variation into populations at a rate that is independent of population size, since the mutation process is density-independent. However, each generation a fraction of the standing variation, $1/2N_e$, is lost via random genetic drift. Provided N_e remains constant, a stochastic equilibrium exists between these opposing forces, leading to an expected equilibrium within-population genetic variance that is directly proportional to N_e and the mutational variance,

$$\sigma_g^2 = 2N_e \left[\sigma_{m0}^2 + \frac{2}{3} (\sigma_{m1}^2 + \sigma_{m2}^2) \right], \qquad\qquad (5.2)$$

where $\sigma_{m0}^2 = 2uE(a^2)$, $\sigma_{m1}^2 = 2uE(a^2k)$, and $\sigma_{m2}^2 = 2uE(a^2k^2)$ (after Lynch and Hill 1986). This simplifies to $2N_e\sigma_{m0}^2$ for mutations with purely additive $(k = 0)$ or purely recessive $(k = -1)$ effects, and to $(14N_e/3)\sigma_{m0}^2$ for completely dominant mutations. All of the variance defined in eq. (5.2) is additive, except for $2N_e\sigma_{m2}^2/3$, which defines the dominance genetic variance. Nonadditive mutations in linkage disequilibrium can cause an inflation of the genetic variance above that defined by eq. (5.2) (Comstock and Robinson 1952), but this is unlikely to be quantitatively significant in the neutral case (Lynch and Hill 1986).

If descendant lines are extracted from an ancestral population and maintained in isolation, their mean phenotypes are expected to drift apart due both to the random fixation of variation from the base population and to independent appearance of new mutations. Provided the base population is in drift-mutation equilibrium, and the descendant lines are maintained at the same effective sizes, the expected between-population variance is simply

$$\sigma_b^2(t) = 2\sigma_{m0}^2 t. \tag{5.3}$$

Thus the variance among population mean genotypic values is expected to increase at the rate $2\sigma_{m0}^2$ per generation. This is a fairly robust result, as it does not depend on the mating system, the population size, the degree of dominance, or the linkage relationships between constituent loci (Lynch and Hill 1986). The asymptotic divergence rate is likely to increase when mutations have epistatic effects (Tachida and Cockerham 1990), although a general model incorporating epistatic effects has not been developed.

Eq. (5.3) can also be obtained readily from the neutral theory of molecular evolution. The expected number of mutational substitutions per locus between two populations isolated for t generations is $2\mu t$. Each substitution causes an expected increment in the between-population variance of $E[(2a - 0)^2]/2 = 2E(a^2)$. Summing over all loci, the expected between-population variance is $n \cdot (2\mu t) \cdot 2E(a^2) = 2\sigma_{m0}^2 t$.

The Lynch-Hill model assumes that the spectrum of mutational effects is independent of the state of the mutating allele. That is, regardless of the initial allelic value or the number of mutations that have occurred at a locus, the expected moments of mutational effects with respect to the parental allele are always $E(a)$, $E(a^2)$, $E(a^2k)$, and $E(a^2k^2)$. With this type of mutation model, there is an infinite number and an infinite range of possible allelic effects.

A rather different perspective was adopted by Cockerham and Tachida (1987), who restricted their attention to an additive model with a finite number of possible allelic states per locus. In the Cockerham-Tachida model, each gene mutates to the ith state at the same rate v_i per generation. With that assumption, the distribution of mutational effects depends completely on the preexisting states of alleles. For example, the effects of alleles at the highest possible state can mutate only in a downward direction, and vice versa.

Cockerham and Tachida's assumption of a limited range of allelic variation leads to predictions that differ from those of Lynch and Hill. In the latter case, the expected genetic variance within populations, $\sigma_g^2 = 2N_e\sigma_{m0}^2$, increases without bound with increasing population size. The Cockerham-Tachida model leads to $\sigma_g^2 \cong 2N_e\sigma_{m0}^2/(1 + 4N_e\mu)$, which converges to $\sigma_{m0}^2/2\mu = nE(a^2)$ as $4N_e\mu \to \infty$. In the Lynch-Hill model, the between-population variance increases indefinitely at the rate $2\sigma_{m0}^2$ per generation, whereas in the Cockerham-Tachida model $\sigma_b^2(t) \cong [1 - (1 - \mu)^{2t}]\sigma_{m0}^2/\mu$, which converges to an equilibrium level of divergence σ_{m0}^2/μ (twice the expected within-population genetic variance) as $t \to \infty$.

   Despite the substantial differences between the assumptions of the two models, their predictions are actually quite similar under a broad range of conditions. When $4N_e\mu \ll 1$, both models yield essentially the same predicted values for σ_g^2, and provided $t \ll 1/2\mu$, both models predict nearly the same between-population divergence, $2\sigma_{m0}^2 t$. Thus, although the Lynch-Hill approach will be adhered to for the remainder of this chapter, it will have little influence on the issues to be discussed, especially if one's interest is confined to variation within and among populations of the same species. If μ is on the order of 10^{-6} to 10^{-5}, the predictions of the two models are essentially the same provided N_e and t are on the order of 10^4 or less. If the assumptions of Lynch and Hill are reasonable, the theory described above will apply to much larger N_e and t as well.

Subdivided Populations

The results given in the previous section apply to the ideal situation in which local populations are panmictic and isolated completely from one another. In nature, however, it is not uncommon for the total population to be fragmented into multiple demes among which there is restricted gene flow. In the following section, we will assume that there is some possible migratory route, either direct or indirect, between all demes. In that case, the between-deme variance has an equilibrium value defined by the balance between the diversifying forces of drift and mutation and the unifying force of migration. The genetic variance expected within the various demes depends on their individual effective sizes as well as on migration rates and routes. Still, some remarkably general results emerge for characters with a purely additive genetic basis. We will consider four types of population subdivision here.

1. THE ISLAND MODEL

General results with respect to population sizes and migration rates were obtained for the case of two populations by Lynch (1988a). When both populations are equivalent in size (N) and migration rate (m, the fraction of the population exchanged prior to mating each generation), the expected within-population variance is approximately $4N\sigma_{m0}^2$. Lande (1992) showed subsequently that this result can be extended to an arbitrary number of demes (d). Provided all demes are equivalent with respect to N and m, and migration events are random among all demes, the expected genetic variance within each deme is $2dN\sigma_{m0}^2$. Thus, with an ideal island structure, the equilibrium within-deme variance is completely independent of the migration rate. Surprisingly, it is the same as if each deme were panmictic with effective size dN. Although the within-deme variance is supplemented each generation by migration, the amount of interdemic divergence (which determines the diversifying effect of a migration event) is inversely proportional to the level of migration. The two effects cancel each other out perfectly.

Lande's (1992) result can be derived, and generalized even further, by using results of gene genealogy theory, an approach that provides the key to analyzing more complicated types of population structure. Slatkin (1987b) and Strobeck (1987) showed that when migratory routes exist between all demes, the mean number of mutational differences between random pairs of neutral genes within a deme (averaged over all demes) is $4N_T\mu$, where $N_T = \sum_{i=1}^{d} N_i$ is the number of individuals summed over all demes. Noting that the expected mean-squared effect of a single mutation is $E(a^2)$, that the contribution of each mutational change to the genetic variance is $E(a^2)/2$, and that $2n$ genes contribute to the character, the average within-deme genetic variance is

$$\sigma_g^2 = 4N_T\mu \cdot 2n \cdot E(a^2)/2 = 2N_T\sigma_{m0}^2. \tag{5.4}$$

Depending on the exact population structure, individual demes may have higher or lower equilibrium variances than this quantity.

The equilibrium between-deme variance under the island model can be obtained by a similar approach. For the ideal case in which all demes are equivalent with respect to N and m, the average number of mutational substitutions that have accumulated at a locus for genes taken from different demes (in excess of those expected on an intrademic basis) is $(d - 1)\mu/m$ (Li 1976). Recalling that the expected contribution of a single substitution to the between-deme variance is $2E(a^2)$, and summing over all n loci,

$$\sigma_b^2 = \frac{(d - 1)\mu}{m} \cdot n \cdot 2E(a^2) = \frac{(d - 1)\sigma_{m0}^2}{m}. \tag{5.5}$$

Lande (1992) obtained the same result via a different route. Thus, under the ideal island model, the between-deme variance depends only on the migration rate per deme $m/(d - 1)$, and the rate of polygenic mutation σ_{m0}^2. As in the case of isolated populations, σ_b^2 is independent of the size of the individual demes. Under this type of population structure, the relative magnitude of interdemic variance is low unless the expected number of immigrants per generation (mN) is less than one. Assuming large d, the ratio of the expected between- and within-deme genetic variances is $1/(2mN)$.

Lande (1992) also investigated the effects of local extinction and colonization in the island model. The number of demes was assumed to be large, and local demes were assumed to go extinct randomly with probability $c \ll 1$ per generation. Following an extinction event, a colonist pool is derived from p random demes, and N_f random members of the colonist pool found the new colony, which is assumed to reach the expected local deme size N in a single generation. Under these conditions, the genetic variance expected within demes is

$$\sigma_g^2 \cong 2dN_e \Lambda \sigma_{m0}^2, \tag{5.6}$$

where $N_e = [(1 - c)/N + c/N_f]^{-1}$ is the effective size of local demes, $\Lambda = 1 - [\lambda/2N_f(2m + \lambda)]$, and $\lambda = c[1 - (1/p)]$. Local extinction and colonization reduce σ_g^2 through a reduction in local effective population sizes via the founder effect ($N_f < N$). If the colonist pool is formed from a single deme ($p = 1$), $\Lambda = 1$, and σ_g^2 is the same as that expected in a panmictic population with effective size dN_e. For $p > 1$, σ_g^2 is smaller than $2dN_e\sigma_{m0}^2$ by a factor between 1 and $1 - (1/N_f)$. The expected equilibrium between-population variance is

$$\sigma_b^2 \cong \frac{(d - 1)\sigma_{m0}^2}{m + (\lambda/2)}. \tag{5.7}$$

If founder pools are extracted from a single deme, $\lambda = 0$, and σ_b^2 is identical to that expected when there is no local extinction (eq. 5.5). When $p > 1$, local colonization and extinction cause a reduction in the divergence between populations since newly founded populations are an average of ancestral populations, and this can be substantial if the extinction probability, $c \cong \lambda$, exceeds $2m$ or so.

2. THE STEPPING-STONE MODEL

The ideal island model assumes that there is an equal probability of migration between all demes. At the opposite extreme is the situation in which migration events occur only between adjacent demes—the stepping-stone model. Only the results for the between-population variance are altered. Consider, for example, a circle of d demes with each deme contributing $mN/2$ migrants per generation to each adjacent deme. When d is large, the circular stepping-stone model results converge on the results for a linear array of populations (Kimura and Weiss 1964). Results in Slatkin (1991) indicate that the average number of mutational substitutions per locus for genes sampled from demes i steps apart is approximately $\mu i(d - i)/m$. Multiplying this quantity by both the number of loci and the phenotypic variance per mutational substitution, the expected genetic variance between demes i steps apart is

$$\sigma_b^2 \cong \frac{i(d - i)\sigma_{m0}^2}{m}. \tag{5.8}$$

Provided $d \gg i$, there is a linear decline in the expected divergence among demes with distance, i.e., $\sigma_b^2 \cong id\sigma_{m0}^2/m$. Note that when $i = 1$ (adjacent populations), eq. (5.8) gives the same result as the island model.

3. ISOLATION BY DISTANCE

A third type of population structure, common in nature, involves isolation by distance, where the population occupies a continuous range but migration is restricted locally. To my knowledge, the theory for this situation has not been worked out formally, but an analogy with the preceding result may give a close approximation to the truth. In the linear stepping-stone model, where there is a probability $m/2$ of migrating either to the right or to the left, i/m can be thought of as the number of steps apart divided by the variance in dispersal distance. Assuming that individuals are distributed and disperse randomly, and letting σ_d^2 be the variance in dispersal distance, the isolation-by-distance analog of this quantity is x/σ_d^2, where x is the distance between individuals in the linear array. This leads to an expected phenotypic variance among individuals separated by distance x of

$$\sigma_b^2 \cong \frac{dx\sigma_{m0}^2}{\sigma_d^2}, \tag{5.9}$$

assuming $d \gg x$, again implying a linear increase in the phenotypic variance among individuals with the geographic distance separating them.

4. HIERARCHICAL STRUCTURE

Recently, Slatkin and Voelm (1991) evaluated the genealogical properties of a population with a hierarchical structure where there are k neighborhoods, each containing d demes. The migration rate between demes within a neighborhood is m_1, while that between neighborhoods is m_2. Applying the methods outlined above to their results, the expected genetic variance within a deme can be shown to be $\sigma_g^2 = 2Nkd\sigma_m^2$. Since Nkd is the total population size, this result is completely compatible with the island-model result. The expected variance among demes within a neighborhood is $k(d-1)\sigma_{m0}^2/m_1$, while that between neighborhoods is $(k-1)\sigma_{m0}^2/m_2$.

The above examples should establish the general procedures for evaluating the expected structure of quantitative-genetic variance of neutral characters with an additive-genetic basis under more complex population structures. Provided there are possible migratory routes between demes, the expected within-population variance is always $2N_T\sigma_{m0}^2$. The variance between any pair of demes is simply the expected number of substitutions per locus (in excess of the expected number within populations) times the quantity $2nE(a^2)$. The expected number of substitutions per locus can be obtained from coalescent theory (Hudson 1991; Slatkin 1993) by multiplying the expected time for random genes from two populations to trace back to the same ancestral populations by 2μ, where the 2 accounts for mutations down both branches. The necessary theory has been developed for many more types of population structure, including two-dimensional stepping-stone models, than discussed above (Maruyama 1970a,b; Slatkin 1991, 1993). Of future interest will be the development of models for subdivided populations that incorporate dominance.

Haploid Sexual Populations

The same general procedures used for the analysis of sexual diploids can be applied to sexual haploid populations, keeping in mind that (1) the initial frequency of a mutant allele is $1/N$ rather than $1/2N$; (2) an expected μN rather than $2\mu N$ mutations arise per locus per generation; (3) there is no dominance; and (4) the ultimate effect of a substitutional change between populations is $E(a)$ rather than $E(2a)$. In the ideal case of random mating within demes, complete isolation between demes, and negligible epistatic effects, the expected means and genetic variances are

$$\bar{g}(t) = \bar{g}(0) + utE(a), \tag{5.10}$$

$$\sigma_g^2 = N_e \sigma_{m0}^2/2, \tag{5.11}$$

$$\sigma_b^2(t) = \sigma_{m0}^2 t/2, \tag{5.12}$$

where σ_{m0}^2 is still defined as $2uE(a^2)$. Thus, compared to a diploid population with similar mutational properties, the mean of a neutral quantitative trait in a haploid population changes at half the rate. The equilibrium genetic variance within haploid sexual populations is about one-quarter or less of that expected under diploidy, and the expected rate of between-population divergence is one-quarter of the diploid expectation.

Asexual Populations

Asexual organisms that reproduce without meiosis pass all of their genes on to their progeny. Consequently, their genomes are functionally equivalent to a single supergene. This greatly simplifies the mathematics needed to derive the neutral model, even allowing the incorporation of epistatic interactions of new mutations, as will be demonstrated below. In the following, it is assumed that the genome is diploid, so that the genomic mutation rate is $2u$, as in the case of sexual diploids. Results for haploid asexuals can be obtained by substituting u for $2u$ in the following expressions.

Consider first the change in the mean phenotype of a clonal population as new mutations accumulate. Each of the N individuals incurs an expected $2u$ new mutations per generation, with fixation probability $1/N$. For asexual organisms, the ultimate effect of a mutational substitution is $E[a(1 + k)]$ rather than $E(2a)$ due to the fact that homozygous mutants cannot be produced by segregation and are highly unlikely to arise by duplicate mutation. The contribution of single-locus effects to the mean phenotype after t generations of mutation accumulation is therefore $2Nut(1/N)E[a(1 + k)] = 2utE[a(1 + k)]$.

The additional contribution from epistatic effects can be obtained in the following way. $n(n - 1)/2$ unique pairs of loci contribute to the trait, and the probability that both members of any pair of loci in the same individual have incurred new mutations by time t is $[1 - \exp(-2ut/n)]^2$. This is approximately $(2ut/n)^2$ provided the genic mutation rate $\mu = u/n \ll 1/2t$. Therefore, assuming n is fairly large so that $n \cong n - 1$, the expected number of pairs of loci with new mutations for both members of the pair by time t is approximately $(2ut)^2/2$. Extending this reasoning, the expected number of groups of x loci that have incurred new mutations at all loci is approximately $(2ut)^x/x!$ provided n is large. Letting $E(a_x)$ represent the expected x-locus interactive effects of new mutations, the dynamics of the expected mean phenotype are then

$$\bar{g}(t) = \bar{g}(0) + 2utE[a(1 + k)] + \frac{(2ut)^2 E(a_2)}{2} + \frac{(2ut)^3 E(a_3)}{6} + \cdots . \quad (5.13)$$

Thus, the mean phenotype is expected to change nonlinearly with time if new mutations interact with one another in a nonadditive fashion. However, the nonlinearities are unlikely to be very prominent if t is much less than the inverse of the genomic mutation rate $(2u)$.

In a clonal population, the total genetic variance need not be subdivided into additive and nonadditive components, since offspring inherit their additive, dominance, and epistatic effects together. Letting σ_m^2 be the rate of input of total mutational variance, and noting that $1/N_e$ of the standing genetic variance is lost by sampling each generation (in comparison to $1/2N_e$ in the case of sexual diploids), the equilibrium level of within-population genetic variance is found to be

$$\sigma_g^2 = N_e \sigma_m^2. \quad (5.14)$$

In the absence of epistatic effects, σ_m^2 is equivalent to the nonsegregational mutational variance of Lynch and Hill (1986), $\sigma_{ns}^2 = 2uE[a^2(1 + k)^2] = \sigma_{m0}^2 + 2\sigma_{m1}^2 + \sigma_{m2}^2$. Thus, comparing eqs. (5.2) and (5.14), we can see that, for equivalent N_e and mutational parameters, σ_g^2 for a neutral character can be as much as twofold higher in a sexual population.

Clonal lines that are derived from a common stem mother t generations in the past are expected to diverge as they independently acquire new mutations. Letting x be the number of mutations acquired by a line at a locus per generation, the expected between-line variance resulting from a cohort of single-locus effects is $n\{\sigma_x^2 E[a^2(1 + k)^2] + \sigma_{a(1+k)}^2 E^2(x)\}$. Since mutations arise in a Poisson fashion under the neutral model, $E(x)$ and σ_x^2 are each equal to $2ut$, and the preceding quantity is very close to $2uE[a^2(1 + k)^2]t$. Therefore, the expected between-line variance resulting from single-locus effects accumulated over t generations is $\sigma_{ns}^2 t$. Comparing this result with eq. (5.3), it can be seen that the rate of clonal divergence will be less than the rate of divergence of sexual populations unless $(2\sigma_{m1}^2 + \sigma_{m2}^2) > \sigma_{m0}^2$. This requires that the average degree of dominance of new mutations exceeds $k = 0.4$.

Computation of the contribution of higher-order (epistatic) effects to σ_b^2 is somewhat complicated by the need to account for the nonindependence of overlapping groups of loci. However, provided the covariance of interactive effects involving overlapping loci is relatively small and $2ut \ll 1$, the x-locus interactive effects of mutations contribute approximately $(2ut)^x E(a_x^2)/x!$ to the between-line variance. Thus, to a good approximation,

$$\sigma_b^2(t) = \sigma_{ns}^2 t + \frac{(2ut)^2 E(a_2^2)}{2} + \frac{(2ut)^3 E(a_3^2)}{6} + \cdots . \tag{5.15}$$

The dynamics of the between-line variance have a form similar to those of the mean phenotype, except that the various terms are functions of the expected squares of mutational effects. If epistatic effects between new mutations are significant, and $2ut$ is not too small, σ_b^2 will increase nonlinearly with time.

Pleiotropic Effects of Mutations

The joint evolution of multiple characters can be constrained if single mutations simultaneously influence multiple traits. The intrinsic pleiotropic effects of mutations are defined by the mutational covariances: $\sigma_{m0}(x,y) = 2uE(a_x a_y)$, $\sigma_{m1}(x,y) = uE[a_x(a_y k_y)] + uE[a_y(a_x k_x)]$, and $\sigma_{m2}(x,y) = 2uE[(a_x k_x)(a_y k_y)]$, where x and y denote two characters, and u is now the expected number of mutations (pleiotropic and nonpleiotropic) per gamete for characters x and y. The expected genetic covariances of neutral characters within and between populations can be obtained by substituting these expressions for the mutational variances in the preceding expressions for σ_g^2 and σ_b^2. For example, for panmictic diploid sexual populations,

$$\sigma_g(x,y) = 2N_e\{\sigma_{m0}(x,y) + \frac{2}{3}[\sigma_{m1}(x,y) + \sigma_{m2}(x,y)]\} \qquad (5.16a)$$

$$\sigma_b(x,y,t) = 2\sigma_{m0}(x,y)t. \qquad (5.16b)$$

Since the between-population variances and covariances are all proportional to t and u, the expected between-population genetic correlation,

$$r_b(x,y) = \frac{\sigma_b(x,y)}{\sqrt{\sigma_b^2(x) \cdot \sigma_b^2(y)}} = \frac{E(a_x a_y)}{\sqrt{E(a_x^2)E(a_y^2)}}, \qquad (5.17)$$

is independent of time and of the mutation rate. The within-population genetic correlation has a slightly more complicated form if there is dominance. However, in the case of additivity, the expected genetic correlations within populations are identical to those between populations (Lande 1979), i.e., $r_w(x,y) = r_b(x,y)$.

Estimation Procedures

The theory presented above provides a qualitative picture of the ways in which mutation and random genetic drift interact to influence patterns of variation for neutral characters. A more quantitative assessment of the issues requires actual estimates of the rate, directionality, and variance of mutational effects. The theory itself provides the basis for most of the estimation procedures. The primary emphasis in this section is on statistical procedures for parameter estimation that can be applied to mutation-accumulation experiments. Issues concerning the standard errors of these estimates and hypothesis testing, some of which are considered in the literature cited (especially Lynch 1988b,c, in prep.; Zeng and Cockerham 1991), are much more complicated and will not be covered in any detail.

Before proceeding, however, it is worth emphasizing that any experimental design with a goal of procuring accurate estimates of mutational parameters needs to involve a very large number of lines, maintained for a large number of generations, with adequate controls run in parallel. Imagine a series of replicate experiments, each involving a different set of L lines, and s individuals sampled per population. For each experiment, an estimate of the average within-population genetic variance is obtained. The coefficient of variation (standard deviation divided by the expectation) of the average within-population genetic variance among experiments is approximately

$$CV(\bar{\sigma}_g^2) = \sqrt{\frac{1}{L}\left(\frac{1}{4N_e u} + \frac{2}{3N_e} + \frac{2}{s}\right)} \qquad (5.18)$$

for populations in drift-mutation equilibrium (Zeng and Cockerham 1991). This expression does not include the sampling variance due to the actual estimation procedure. It only accounts for the random variation that arises among replicate populations at any one point in time and for the limited sample size available to the investigator. Usually, this sort of error is not even considered in the analysis of data. Yet in most mutation-accumulation experiments N is on the order of 1 to 3, so $(1/4N_e u)$ + $(2/3N_e)$ will be on the order of 1 or so. Thus, even if s is very high, the number of lines L needs to be on the order of 100 (and perhaps much more) to achieve an accuracy of $CV(\bar{\sigma}_g^2) = 0.1$.

The coefficient of variation of the between-population variance is approximately

$$CV(\sigma_b^2) = \sqrt{\frac{2}{L-1}} \qquad (5.19)$$

provided the number of loci is fairly large (Lynch and Hill 1986; Zeng and Cockerham 1991). Again since this expression estimates only the coefficient of variation of the σ_b^2 realized in a set of replicate experiments, each involving L lines, and does not include the additional variation due to the limitations on statistical analysis, it is clear that L needs to be at least 200 if the desired accuracy is on the order of $CV(\sigma_b^2) = 0.1$.

All of the techniques outlined below rely on observed trends in the mean and variance of a trait. Since the quantities to be estimated are often very small, changes in environmental conditions during the course of an experiment can cause substantial bias in parameter estimation. One way to guard against this problem is to utilize a genetically constant control (such as seed stored from the base population) throughout the course of the experiment, and to treat the control as a covariate to remove any environmental trends prior to analysis (Muir 1986; Lynch 1988c). An alternative procedure is to store gametes or zygotes from the experimental lines at various stages of the experiment and to analyze them simultaneously in the same environmental setting at a later date. It will be assumed below that efforts along these lines have been made prior to analysis.

The expressions given below are for diploid sexual populations. However, only slight modifications by factors of two, which follow from the theory given above, need to be made for haploid or asexual organisms.

Trends in the Mean and Variance

The simplest approach to estimating the properties of polygenic mutation is motivated by the theoretical trends of the mean phenotype and the variance among isolated lines. As outlined above, provided the base population is in drift-mutation equilibrium and the descendant lines are maintained in isolation at the same effective size, both the mean and the variance are expected to change linearly through time as mutations accumulate in a neutral fashion. Not all mutations are neutral, of course, but in small populations non-neutral mutations will be overwhelmed by random genetic drift. Alleles behave in an effectively neutral fashion provided the absolute value of the selection coefficient is less than $1/4N_e$ (Kimura and Ohta 1971). Therefore, to insure that estimates of mutational properties are as nonbiased as possible, the usual strategy in a mutation-accumulation experiment is to start with a highly inbred base population and to maintain the descendant lines at a very small size.

In many plants, a base population can consist of a single individual obtained after many generations of self-fertilization and single-seed descent. The descendant lines, ideally all derived from the same parent, can then be maintained in the same manner. For species with separate sexes, the same strategy can be employed through full-sib mating. In asexual species, the base population can consist of a single individual whose derived lines are maintained by single-seed descent. In *Drosophila melanogaster*, for which chromosomes with visible markers are available, single chromosomes can be passed on in a clonal manner through males, which do not exhibit recombination (Mukai 1979).

From eq. (5.1) the mean phenotype is expected to change at the rate $2uE(a)$ per generation. An estimate of this quantity, R_m, can be obtained by calculating the slope of the regression of the mean phenotype of the experimental lines on time. From eq. (5.13) the expected value of the between-line variance is $2\sigma_{m0}^2 t$. Thus, an estimate, V_{m0}, of the mutational variance $\sigma_{m0}^2 = 2uE(a^2)$ can be obtained as one-half the slope of a regression of the between-line variance on time. Typically, between-line variance estimates are obtained by use of analysis of variance techniques. Ideally, regression analysis of temporal trends in the mean and variance should involve a generalized least-squares approach in order to account for the nonindependence of data (Mackay et al. 1992; Lynch, in prep.). In order to yield a dimensionless parameter so that different characters can be compared, V_{m0} is usually divided by the environmental variance of the trait, σ_E^2. The mutational heritability $h_m^2 = \sigma_{m0}^2/\sigma_E^2$ is equivalent to the gain in heritability that occurs in the first generation of mutation accumulation in a pure line for a character with an additive genetic basis.

These general procedures, as well as observations on the response of inbred lines to selection, have yielded estimates of h^2_m for a diversity of morphological and life-history characters in plants and animals (Lynch 1988b). These range from about 10^{-4} to 10^{-2}, with most estimates being $\geq 10^{-3}$ and no obvious patterns with respect to character type or taxonomic group. A good deal of the variation among observed values of h^2_m is undoubtedly due to sampling error, but it is known that h^2_m can be elevated beyond 10^{-2} by transposable-element activity in *Drosophila melanogaster* (Yukuhiro et al. 1985; Mackay 1989). The high average value of $h^2_m = 1.6 \times 10^{-2}$ for skeletal characters in mice reported by Lynch (1988b) is identical to a recent estimate for growth rate (Keightley and Hill 1992). Lande's (1975) suggestion that $h^2_m \cong 10^{-3}$ for bristle numbers in nondysgenic lines of *Drosophila* has also stood up to subsequent scrutiny (Lynch 1988b; Caballero et al. 1991), and is essentially the same as a recent estimate for alcohol vapor resistance (Weber and Diggins 1990).

Little attention has been given to the directional effects of spontaneous polygenic mutations, but R_m is often approximately zero (Lynch 1988b). However, mutations for viability, to be discussed below, are biased strongly in the negative direction (Mukai 1979).

Line-Cross Analysis

As the isolated lines in a mutation-accumulation experiment diverge, they will develop inbreeding depression with respect to each other if the mutations have nonadditive effects. Information on the average degree of dominance can be obtained only by hybridizing the lines and comparing the resultant mean phenotype ($\bar{z}_x$) with that of the parental lines ($\bar{z}_s$). Assuming that different lines acquire mutations at unique loci, the F_1 progeny will be heterozygous at all loci that have acquired a mutation in one or the other parent. Starting from a base population in drift-mutation equilibrium, the expected difference between the mean phenotype of hybrid progeny and that of their contemporary parents is $2uE(ak)t$ (Lynch, in prep.). Thus, $2uE(ak)$ can be estimated in a single experiment by dividing ($\bar{z}_x - \bar{z}_s$) by the number of generations of divergence, or from a temporal series of experiments by evaluating the rate of change of ($\bar{z}_x - \bar{z}_s$) across generations. Below, I refer to such an estimate as D_m.

When the members of a set of L divergent lines are crossed in all possible combinations in a diallel experiment, it is also possible to estimate the three mutational variances σ^2_{m0}, σ^2_{m1}, and σ^2_{m2}. Again assuming the mutation-accumulation experiment has been initiated from an equilibrium base population t generations in the past, the expected variance among the observed mean phenotypes for the L pure-bred families is

$$\sigma_s^2(t) = 2\sigma_{m0}^2 t + \sigma_{e,s}^2. \tag{5.20}$$

This is simply the expected between-line genetic variance plus the sampling variance of the individual means, where the latter is the average variance among individuals within pure-bred families divided by the number of individuals measured per family.

Two additional statistics can be computed from the observed family means in a diallel. For each of the L lines, marginal means can be calculated. Each marginal mean involves a cross of the line to itself and to the $L - 1$ remaining lines. σ_r^2 is the variance among the L marginal means. Within each line, the variance of the L family means around the marginal mean can also be computed; the expected average of these L estimates is σ_f^2. Following procedures outlined in Hayman (1954) and Mather and Jinks (1982), and assuming that mutations in different lines occur at unique loci, these two parameters can be described in terms of causal components,

$$\sigma_r^2 = \frac{\sigma_{m0}^2 t}{2} + \phi \sigma_{m1}^2 t + \frac{\phi^2 \sigma_{m2}^2 t}{2}, \tag{5.21a}$$

$$\sigma_f^2 = \frac{\sigma_{m0}^2 t}{2} + \phi \sigma_{m1}^2 t + \frac{\sigma_{m2}^2 t}{2} + \frac{(L-1)\sigma_{e,c}^2 + \sigma_{e,s}^2}{L}, \tag{5.21b}$$

where $\phi = 1 - (2/L)$, and $\sigma_{e,c}^2$ is the sampling variance of the mean for a cross-bred family (not necessarily the same as $\sigma_{e,s}^2$).

Provided $L > 2$, the three mutational parameters can be estimated by substituting estimates of the parameters σ_s^2, σ_r^2, σ_f^2, $\sigma_{e,s}^2$, and $\sigma_{e,c}^2$ into the preceding expressions and solving. Letting V denote a variance estimate after the environmental contribution has been eliminated,

$$V_{m0} = \frac{V_s}{2t}, \tag{5.22a}$$

$$V_{m1} = \frac{1}{t\phi}\left(V_f - 4V_s - \frac{V_f - V_r}{1 - \phi^2}\right), \tag{5.22b}$$

$$V_{m2} = \frac{2(V_f - V_r)}{t(1 - \phi^2)}. \tag{5.22c}$$

The ratios V_{m1}/V_{m0} and V_{m2}/V_{m0} have expected values equal to approximately $E(a^2 k)/E(a^2)$ and $E(a^2 k^2)/E(a^2)$ and hence provide estimates of the average degree of dominance k and the average value of k^2, with each mutation being weighted by its squared effect a^2.

Diallel analysis can also be used to test for the epistatic effects of mutations. For each line, the covariance of the L observed means with the means of the L parallel pure breds, $C_{f,s}$, can be computed. These L covariances can then be regressed on the associated variances among means within rows, i.e., the line-specific V_f. In the absence of epistasis, the expected slope of this regression is equal to 1 (Mather and Jinks 1982). Provided that is the case, the expected intercept is $(\sigma_{m0}^2 - \sigma_{m2}^2)/2$. Thus a positive intercept is expected under partial dominance $(\overline{k^2} < 1)$, a zero intercept with complete dominance $(\overline{k^2} = 1)$, and a negative intercept with overdominance $(\overline{k^2} > 1)$.

Ideally, line-cross analyses ought to be done by performing reciprocal crosses between lines to control for maternal effects, and multiple crosses should be performed with each combination of parental lines. It is often impractical to perform individual diallels with large numbers of lines, and there are advantages to partitioning the lines into a series of blocks. If the lines are allocated to several separate experiments, the individual estimates of the mutational parameters can be pooled to obtain average estimates as well as their standard errors.

The Bateman-Mukai Technique

All of the quantities that are directly observable from mutation-accumulation experiments (R_m, V_{m0}, V_{m1}, V_{m2}, and D_m) are composite functions of u, $E(a)$, and higher-order moments of a and k. Various aspects of evolutionary theory require a knowledge of these subsidiary components (Crow and Simmons 1983; Turelli 1984).

Bateman (1959) first suggested how simple algebraic functions of R_m and V_{m0} can be used to estimate one-sided limits for u and $E(a)$, and his results were extended and generalized to account for sampling error by Lynch (in prep.). Letting $C_R = \text{Var}(R_m)/R_m^2$ and $C_V = \text{Var}(V_{m0})/V_{m0}^2$, where $\text{Var}(R_m)$ and $\text{Var}(V_{m0})$ are the sampling variances of R_m and V_{m0}, be the squared sampling coefficients of variation, then a downwardly biased estimate of the gametic mutation rate is

$$\hat{u}_{min} = \frac{R_m^2(1 - C_R)}{V_{m0}(2 + C_V)}. \tag{5.23}$$

The parameter $u_{min} = [2\mu E(a)]^2/2\sigma_{m0}^2$ is actually equal to $u/(1 + \theta)$, where $\theta = \sigma_a^2/E^2(a)$ is the squared coefficient of variation of mutational effects, with σ_a^2 being the variance of mutational effects. Thus the degree to which $\hat{u}_{min}$ underestimates the true mutation rate depends on the extent of variation among mutational effects. Estimates of the upper bounds on the average and the mean squared mutational effects are

$$\hat{a}_{max} = \frac{V_{m0}}{R_m^2(1 + C_R)}, \tag{5.24a}$$

$$\hat{a}_{max}^2 = \left(\frac{V_{m0}}{R_m}\right)^2 \left(\frac{1}{1 + C_V + 3C_R}\right). \tag{5.24b}$$

Since $\hat{a}_{max}$ actually provides an unbiased estimate of $(1 + \theta)E(a)$, and $\hat{a}_{max}^2$ of $(1 + \theta)E(a^2)$, the degree to which they overestimate $E(a)$ and $E(a^2)$ again depends on θ.

Care should be taken in the interpretation of these estimators. They provide *estimates* of the bounds on u, $E(a)$, and $E(a^2)$. They are not the bounds themselves, and without information on σ_a^2, it is difficult to say how far the expectations of the estimates are from the true parametric values.

Finally, following up on the relationship pointed out in the preceding section, an estimator for the average degree of dominance is

$$\hat{k} = \frac{D_m}{R_m(1 + C_R)}. \tag{5.25}$$

$\hat{k}$ actually provides an unbiased estimate of $E(k)\{1 + \sigma_{a,k}/[E(a)E(k)]\}$, where $\sigma_{a,k}$ is the covariance of the effects a and k in new mutations. $\hat{k}$ can be an upwardly or downwardly biased estimate of $E(k)$ depending on the signs of $\sigma_{a,k}$, $E(a)$, and $E(k)$. However, when the values of a and k for mutations are uncorrelated, $\hat{k}$ estimates $E(k)$, the average degree of dominance.

Procedures similar to those outlined above have been used extensively by Mukai and colleagues to estimate the properties of viability mutations in *Drosophila melanogaster*. The results of numerous studies, summarized in Mukai (1979, 1980) and Crow and Simmons (1983), lead to the average estimates: $\hat{u}_{min} = 0.3$, $\hat{a}_{max} = 0.0125$, and $\hat{k} = -0.2$. That is, zygotes appear to acquire an average of at least 0.6 new viability mutations per generation. These mutations are only slightly recessive, and the average effect of an individual mutation in the homozygous condition is no greater than a 2.5% reduction in viability. These results are somewhat conservative in that chromosomes with highly deleterious mutations were excluded from the final analysis. Lethal mutations arise at about 1% of the rate of nonlethals.

Houle et al. (1992) recently extended this line of work in an evaluation of mutations for total fitness in mutation-accumulation lines of *D. melanogaster*, obtaining estimates of $\hat{u}_{min} \cong 0.25$ and $\hat{a}_{max} \cong 0.03$. The fact that both types of analyses yield similar estimates for $\hat{u}_{min}$ suggests that the same mutations may be influencing viability and total fitness. It would then follow that the higher value of $\hat{a}_{max}$ for total fitness is due to the pleiotropic effects of mutations on multiple-fitness components.

Discussion

Substantial progress has been made recently in the development of neutral models for phenotypic evolution. The procurement of data on the properties of polygenic mutation has been much slower in coming, and we are unlikely to advance very far beyond the current state of understanding without the performance of some very large experiments involving inbred lines and tens to hundreds of thousands of individuals. Although we have a rough understanding of the magnitude of mutational heritability, the range of available estimates is about two orders of magnitude, and it is not known how much of this range is simply due to estimation error. Except for some rough estimates for fitness in *D. melanogaster*, we know essentially nothing about the gametic mutation rate or the distribution of mutational effects. Although that level of information is not absolutely essential for our understanding of patterns of neutral evolution, it is critical to the development of theory for the evolution of characters under natural selection (Turelli 1984), and for understanding of patterns of mutation load (Crow 1970), inbreeding depression (Morton et al. 1956), and extinction via mutation accumulation in small populations (Lynch and Gabriel 1991). Also central to these problems are the degree of epistatic and pleiotropic effects of polygenic mutations (especially on primary fitness components). Except for the observation in *D. melanogaster* that mutations for bristle number with large enough effects to be individually discernible often have negative pleiotropic effects on fitness (Mackay 1989), almost nothing is known about these issues.

The neutral models presented in this paper are concerned with characters expressed at the phenotypic level, i.e., behavioral, morphological, physiological, and life-history traits. These are, of course, precisely the characters on which natural selection is most likely to operate. The neutral model provides a null hypothesis for explicitly testing this supposition, by predicting the genetic variance within populations of known size in the absence of selection, and by predicting the extent of divergence of population mean phenotypes when the degree of population isolation is known. Statistical tests for selection using neutral divergence as a null hypothesis, and applications of them, are outlined in Lande (1976), Turelli et al. (1988), and Lynch (1990).

Finally, it is worth noting that the neutral theory provides a useful tool for conservation genetics. In many programs for the management of endangered species, two of the major goals are the maintenance of genetic variation and the minimization of evolutionary change of the mean phenotype. As a consequence of the provisioning of food and mates and protection against predators and pathogens, natural selection is often relaxed almost entirely for species maintained in zoological parks or botanical gardens. In that type of setting, random genetic drift and polygenic mutation become the dominant factors determining the evolutionary dynamics of the means, variances, and covariances of quantitative characters. Thus the theory outlined above can be used to predict the consequences of policy changes that modify the effective population size (N_e) of the captive population.

A general conclusion that emerges from the neutral theory is that the amount of genetic variation that can be maintained for a quantitative trait is directly proportional to N_e. On the other hand, the rate at which the mean phenotype changes due to directional mutational effects is independent of N_e, and the same is true for the potential rate of drift of the mean phenotype of the captive isolate away from the ancestral state. Thus, while the payoff of a doubling of N_e is an approximate doubling of the amount of genetic variance within a managed population, unless selection is imposed no amount of population expansion will stabilize the mean phenotype.

Acknowledgments

I am very grateful to W. G. Hill and R. Lande, who have been very useful sources of advice, ideas, and inspiration. M. Slatkin provided substantial advice as I tried to understand how the theory for subdivided populations can be developed. Many thanks to E. Martins for helpful comments. This work has been supported by NSF grants BSR 89-11038 and BSR 9024977, and PHS grant GM36827-01.

References

Bateman, A. J. 1959. The viability of near-normal irradiated chromosomes. *Internat. J. Rad. Biol.* 1: 170–180.

Bürger, R., G. Wagner, and F. Stettinger. 1989. How much heritable variation can be maintained in finite populations by mutation-selection balance? *Evolution* 43: 1748–1766.

Caballero, A., M. A. Toro, and C. López-Fanjul. 1991. The response to artificial selection from new mutations in *Drosophila melanogaster. Genetics* 127: 89–102.

Chakraborty, R., and M. Nei. 1982. Genetic differentiation of quantitative characters between populations or species. I. Mutation and random genetic drift. *Genet. Res.* 39: 303–314.

Clayton, G. A., and A. Robertson. 1955. Mutation and quantitative variation. *Amer. Nat.* 89: 151–158.

Cockerham, C. C., and H. Tachida. 1987. Evolution and maintenance of quantitative genetic variation by mutations. *Proc. Natl. Acad. Sci. USA* 84: 6205–6209.

Comstock, R. E., and H. F. Robinson. 1952. Estimation of average dominance of genes. In *Heterosis*, ed. J. Gowen, pp. 494–516. Iowa State University Press, Ames, Iowa.

Crow, J. F. 1970. Genetic loads and the cost of natural selection. In *Mathematical Topics in Population Genetics*, ed. K. Kojima, pp. 128–177. Springer-Verlag, Berlin.

Crow, J. F., and M. J. Simmons. 1983. The mutation load in *Drosophila*. In *The Genetics and Biology of Drosophila*, ed. M. Ashburner, H. L. Carson, and J. N. Thompson, Jr., vol. 3c, pp. 2–35. Academic Press, New York.

Foley, P. 1992. Small population genetic variability at loci under stabilizing selection. *Evolution* 46: 763–774.

Gillespie, J. 1991. *The Origin and Maintenance of Genetic Variance*. Oxford University Press, New York, NY.

Hayman, B. I. 1954. The theory and analysis of diallel crosses. *Genetics* 39: 789–809.

Houle, D. 1989. The maintenance of polygenic variation in finite populations. *Evolution* 43: 1767–1780.

Houle, D., D. K. Hoffmaster, S. Assimacopoulos, and B. Charlesworth. 1992. The genomic mutation rate for fitness in *Drosophila*. *Nature* 359: 58–60.

Hudson, R. 1991. Gene genealogies and the coalescent process. *Oxford Surv. Evol. Biol.* 7: 1–44.

Keightley, P. D., and W. G. Hill. 1992. Quantitative genetic variation in body size of mice from new mutations. *Genetics* 131: 693–700.

Kimura, M. 1968. Evolutionary rate at the molecular level. *Nature* 217: 624–626.

Kimura, M. 1983. *The Neutral Theory of Molecular Evolution*. Cambridge University Press, Cambridge, U.K.

Kimura, M., and T. Ohta. 1971. On the rate of molecular evolution. *J. Molec. Evol.* 1: 1–17.

Kimura, M., and G. Weiss. 1964. The stepping-stone model of population structure and the decrease of genetic correlation with distance. *Genetics* 49: 561–576.

Lande, R. 1975. The maintenance of genetic variation by mutation in a quantitative character with linked loci. *Genet. Res.* 26: 221–235.

Lande, R. 1976. Natural selection and random genetic drift in phenotypic evolution. *Evolution* 30: 314–334.

Lande, R. 1979. Quantitative genetic analysis of multivariate evolution, applied to brain: body size allometry. *Evolution* 33: 402–416.

Lande, R. 1992. Neutral theory of quantitative genetic variance in an island model with local extinction and colonization. *Evolution* 46: 381–389.

Li, W.-H. 1976. Distribution of nucleotide differences between two randomly chosen cistrons in a subdivided population: The finite island model. *Theor. Pop. Biol.* 10: 303–308.

Lynch, M. 1988a. The divergence of neutral quantitative characters among partially isolated populations. *Evolution* 42: 455–466.

Lynch, M. 1988b. The rate of polygenic mutation. *Genet. Res.* 51: 137–148.

Lynch, M. 1988c. Design and analysis of experiments on random drift and inbreeding depression. *Genetics* 120: 791–807.

Lynch, M. 1990. The rate of morphological evolution in mammals from the standpoint of the neutral expectation. *Amer. Nat.* 136: 727–741.

Lynch, M., and W. Gabriel. 1991. Mutation load and the survival of small populations. *Evolution* 44: 1725–1737.

Lynch, M., and W. G. Hill. 1986. Phenotypic evolution by neutral mutation. *Evolution* 40: 915–935.

Mackay, T.F.C. 1989. Mutation and the origin of quantitative variation. In *Evolution and Animal Breeding*, ed. W. G. Hill and T.F.C. Mackay, pp. 113–119. CAB International, Wallingford, U.K.

Mackay, T.F.C., R. F. Lyman, M. S. Jackson, C. Terzian, and W. G. Hill. 1992. Polygenic mutation in *Drosophila melanogaster*: Estimates from divergence among inbred strains. *Evolution* 46: 300–316.

Maruyama, T. 1970a. Effective number of alleles in a subdivided population. *Theor. Pop. Biol.* 1: 273–306.

Maruyama, T. 1970b. Stepping-stone models of finite length. *Adv. Appl. Prob.* 2: 229–258.

Mather, K., and J. L. Jinks. 1982. *Biometrical Genetics*. Chapman and Hall, New York.

Morton, N. E., J. F. Crow, and H. J. Muller. 1956. An estimate of the mutational damage in man from data on consanguineous marriages. *Proc. Natl. Acad. Sci. USA* 42: 855–863.

Muir, W. M. 1986. Estimation of response to selection and utilization of control populations for additional information and accuracy. *Biometrics* 42: 381–391.

Mukai, T. 1979. Polygenic mutation. In *Quantitative Genetic Variation*, ed. J. N. Thompson, Jr., and J. M. Thoday, pp. 177–196. Academic Press, New York.

Mukai, T. 1980. The genetic structure of natural populations of *Drosophila melanogaster*. XIV. Effects of the incomplete dominance of the *IN(2LR)SM1 (Cy)* chromosome on the estimates of various genetic parameters. *Genetics* 94: 169–184.

Slatkin, M. 1987a. Heritable variation and heterozygosity under a balance between mutations and stabilizing selection. *Genet. Res.* 50: 53–62.

Slatkin, M. 1987b. The average number of sites separating DNA sequences drawn from a subdivided population. *Theor. Pop. Biol.* 32: 42–49.

Slatkin, M. 1991. Inbreeding coefficients and coalescence times. *Genet. Res.* 58: 167–175.

Slatkin, M. 1993. Isolation by distance in equilibrium and nonequilibrium populations. *Evolution* 46 (in press).

Slatkin, M., and L. Voelm. 1991. F_{ST} in a hierarchical island model. *Genetics* 127: 627–629.

Strobeck, C. 1987. Average number of nucleotide differences in a sample from a single subpopulation: A test for population subdivision. *Genetics* 117: 149–153.

Tachida, H., and C. C. Cockerham. 1990. Evolution of neutral quantitative characters with gene interaction and mutation, In *Population Biology of Genes and Molecules*, ed. N. Takahata and J. F. Crow, pp. 233–249. Baifukan, Tokyo.

Turelli, M. 1984. Heritable genetic variation via mutation-selection balance: Lerch's zeta meets the abdominal bristle. *Theor. Pop. Biol.* 25: 138–193.

Turelli, M., J. H. Gillespie, and R. Lande. 1988. Rate tests for selection on quantitative characters during macroevolution and microevolution. *Evolution* 42: 1085–1089.

Weber, K. E., and L. T. Diggins. 1990. Increased selection response in larger populations. II. Selection for ethanol vapor resistance in *Drosophila melanogaster* at two population sizes. *Genetics* 125: 585–597.

Yukuhiro, K., K. Harada, and T. Mukai. 1985. Viability mutations induced by the P elements in *Drosophila melanogaster*. *Jap. J. Genet.* 60: 307–334.

Zeng, Z-B., and C. C. Cockerham. 1991. Variance of neutral genetic variances within and between populations for a quantitative character. *Genetics* 129: 535–553.

6

Evolutionary Genetics of *Daphnia*

(with Ken Spitze)

Introduction

As a science, ecological genetics has developed in a rather fragmented fashion. Numerous studies on a diversity of species have evaluated the levels of molecular variation existing within and among populations as well as among species (for recent reviews, see Kimura 1983; Nei 1987; Selander et al. 1991). Most of these studies have assumed, either implicitly or explicitly, that the molecular markers under consideration (usually allozymes or restriction fragment length polymorphisms) are neutral enough to allow their use as tools for estimating aspects of population structure, divergence times, etc. (Slatkin, chaps. 1, 2, this volume).

Recently, a great deal of attention has also been focused on the estimation of levels of variation and covariation for polygenic characters in natural populations, using quantitative-genetic techniques that have been established firmly in the realm of plant and animal breeding programs (Travis, chaps. 9, 10; Via, chaps. 3, 4, this volume). Usually, these types of studies have been performed with populations for which molecular data are not available, so our understanding of the relationships between patterns of evolution at the two levels is essentially undeveloped.

Over the past decade or so, our laboratory has been examining genetic variation at the molecular and quantitative-trait levels at both the intraspecific (within and among populations) and interspecific levels. This work utilizes planktonic microcrustaceans in the genus *Daphnia* as a model system. We are still a long way from a full understanding of the evolutionary genetic properties of the genus, but enough progress has been made in some areas that it now seems worthwhile to attempt a synthesis of the results. This chapter summarizes some of the more interesting patterns that have begun to emerge.

Although some *Daphnia* have dropped the sexual phase of the life cycle (Hebert 1981; Hebert et al. 1988), most reproduce by cyclical parthenogenesis, and our attention will be focused entirely on populations with that type of life cycle. Cyclically parthenogenetic *Daphnia* normally reproduce by asexual means so long as the environment remains favorable. When conditions deteriorate, males are produced, as are haploid eggs, which when fertilized give rise to diapausing embryos (ephippia). For ponds that dry completely, annual recruitment is entirely from these sexually reproduced eggs. However, populations inhabiting permanent lakes have the potential to reproduce indefinitely by parthenogenesis.

Evolution at the Molecular Level

Since the pioneering studies of Hebert (1974a,b), dozens of isozyme surveys have been performed on *Daphnia* populations. Most of these studies have involved ten or fewer loci, and attention has often been restricted to those loci that exhibit the greatest degree of polymorphism. Thus it is still somewhat difficult to make precise statements as to the average levels of biochemical diversity within and among populations. However, some generalizations are now possible.

Our surveys of three species of cyclically parthenogenetic *Daphnia* are based on nine to twelve loci selected only for the ease of resolution of their banding patterns on gels. The observed allele frequencies have been used to estimate v_w, the average gene diversity within populations (Nei 1987). This statistic is equivalent to the average (over all loci) heterozygosity that would be observed in the progeny if the sampled population were to undergo random mating. Being based on Hardy-Weinberg expectations, v_w provides a useful measure of isozyme variation that is unbiased by aspects of population structure, local inbreeding, clonal selection, and so forth. For seven central Illinois *D. pulex* populations, we estimate the average v_w to be 0.11 (0.05). (Throughout, numbers in parentheses are standard errors.) An identical estimate has been obtained for nine *D. obtusa* populations in the same general region, whereas the average v_w for eight western Oregon populations of *D. pulicaria* is 0.15 (0.07).

For Czechoslovakian populations of *D. magna*, *D. longispina*, and *D. galeata*, using the six-locus survey of Hebert et al. (1989b), we estimate v_w to be 0.12 (0.06), 0.13 (0.07), and 0.17 (0.10), respectively. Benzie's (1986a) extensive survey of Australian *D. cephalata* and *D. carinata* populations yields rather higher v_w estimates—0.16 (0.06) and 0.21 (0.07)— whereas Korpelainen (1986a) obtains estimates ranging from 0.04 to 0.09 for four European species of *Daphnia*. Taken as a group, these data suggest that *Daphnia* harbor substantial levels of variation at the level of nuclear genes—an average individual appears to be heterozygous at 5% to 20% of its protein-coding loci. These could very well be underestimates since it is well known that protein electrophoresis is unable to discriminate a large fraction of actual variants. To resolve this issue, studies need to be performed at the DNA level.

Hebert (1974a,b) pointed out an interesting feature of the temporal dynamics of genotype frequencies in British populations of *D. magna*. Populations inhabiting intermittent ponds exhibited fairly stable genotype frequencies within and between years. The genotype frequencies were generally in good agreement with Hardy-Weinberg expectations, and gametic-phase disequilibria between loci were not observed. In contrast, genotype frequencies in permanent populations exhibited dramatic temporal instabilities, often deviating far from Hardy-Weinberg expectations and showing marked gametic-phase disequilibria. The observed changes did not repeat themselves annually; rather, the dominant multilocus genotypes varied from year to year. These observations were corroborated by Young (1979a,b), and similar results have been noted in permanent-pond populations of *D. pulex* (Lynch 1983; Weider 1985) and permanent-lake populations of *D. pulicaria* (Lynch et al., in prep.).

Variation in the reproductive mode provides a potential explanation for the dramatic differences in population structure in the two types of environments (Hebert 1974a,b). In intermittent environments, populations are transient, and sexual reproduction is enforced on an annual basis. Recombination every few generations produces an enormous number of new clones each year, and a short growing season does not provide sufficient time for selection to advance appreciably a small number of multilocus genotypes. On the other hand, in permanent environments, periods of purely clonal reproduction can extend for a dozen or more generations. If uninterrupted by significant recruitment of new clones from resting eggs, this can provide ample time for a small fraction of the initial pool of clones to come to dominance (Spitze 1991). Once the total number of clones has been reduced to a small number, there will be a high probability of the chance development of gametic-phase disequilibria, and the multilocus genotypes that are advanced will be simply those that are associated fortuitously with the most favorable clones at the time. A repeatable seasonal cycle of multilocus genotypes is not to be expected if a population is experiencing a low level of recruitment from sexually produced eggs, since that would occasionally produce new associations between the electrophoretic markers and the genes underlying the selected characters.

Unfortunately, the general patterns outlined above have not held up to close scrutiny in other taxa. For example, some large-lake populations of *D. cucullata* and *D. galeata* exhibit a high degree of temporal stability in genotype frequencies with few significant deviations from Hardy-Weinberg expectations (Mort and Wolf 1985, 1986). On the other hand, Finnish rock-pool populations of *D. magna* and *D. pulex* are characterized by large temporal fluctuations of genotype frequencies (Korpelainen 1986b,c). Taken at face value, these discrepancies seem to be inconsistent with Hebert's hypothesis, but they do not necessarily rule it out. Although the rarity of males and ephippial (sexual) females in large-lake populations has led to the general feeling that sexual reproduction is uncommon, it is possible that on a lakewide basis there is enough sexual recruitment to keep the populations in a state similar to that of Hebert's intermittent *D. magna* populations. It is also conceivable that the annual recruitment of sexually produced propagules is quite low in the small rock pools studied by Korpelainen. If that were the case, pronounced gametic-phase disequilibria might arise even in a relatively short growing season.

When isozyme frequencies are available for multiple populations of the same species, it is possible to partition the total gene diversity into within- and between-population components, v_w and v_b (Nei 1987). v_b is equivalent to the expected gene diversity (heterozygosity assuming random mating) between populations that is in excess of that within populations. The index

$$G_{st} = \frac{v_b}{v_w + v_b},$$

which scales from 0 to 1, provides a useful index of population subdivision.

For our surveys of *D. pulex*, *D. pulicaria*, and *D. obtusa*, $G_{st} = 0.18$ (0.07), 0.31 (0.08), and 0.29 (0.05), respectively. These results do not necessarily imply that *D. pulex* has an unusually low degree of population subdivision. The ponds surveyed for this species are no more than 80 km apart, whereas the maximum distances between the *D. pulicaria* and *D. obtusa* populations are 240 and 560 km, respectively. So, in principle, the differences in G_{st} are compatible with an isolation-by-distance hypothesis. Indeed, for *D. pulex* we have shown that G_{st} increases from approximately 0.05 for populations 1 km apart to 0.3 for populations separated by 1000 km (Crease et al. 1990). Korpelainen (1984) has also observed an isolation-by-distance relationship in *D. magna*. Figure 6.1 summarizes data from surveys of a number of *Daphnia* species. Although there is considerable noise in the relationship, it can be seen that in general there is a slow increase in G_{st} with distance.

Provided the allelic variants are effectively neutral, data on spatial and temporal variation of isozyme frequencies can be used to estimate aspects of population structure such as effective population size, migration rates, and so on (Nei 1987; Weir 1990). Although such analyses are now quite common in the literature, they generally have been carried out in the absence of any direct evidence that the influence of selection on the isozyme variants is in fact negligible. Since a large number of *Daphnia* surveys involve long (up to eight-year) temporal sequences of data, it has been possible to employ statistical procedures to evaluate this assumption

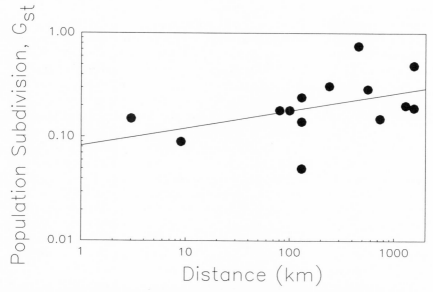

FIG. 6.1. The relationship between the degree of population subdivision and maximum distance among sites involved in a survey for a variety of *Daphnia* species. In addition to our own results, data are given from Korpelainen (1984, 1986a), Benzie (1986a), Mort and Wolf (1986), and Hebert et al. (1989c).

(Lynch 1987). Analyses from several populations lead to the conclusion that the isozyme variants are quasineutral. In no case does the long-term average selection coefficient on an isozyme variant deviate significantly from zero, but on specific dates the observed changes in allele frequencies often exceed the expectations resulting from sampling error alone. Thus the data suggest that periods of positive selection on the isozyme variants are balanced roughly with periods of negative selection. The variants are neutral with respect to each other in the long term but not in the short term.

It has long been known that fluctuating selection of this sort leads to a driftlike phenomenon (Wright 1948; Kimura 1954), causing gene frequencies to diverge among isolated populations more rapidly than expected under neutrality and also influencing the level of heterozygosity that can be maintained within populations. Therefore, the use of isozyme variants to infer properties of population structure in *Daphnia* needs to be tempered with a great deal of caution.

The selection analyses in Lynch (1987) provide insight into some previously puzzling observations. For example, the permanent *D. magna* populations of Hebert (1974a) often exhibited dramatically different gene frequencies even when only meters apart, whereas the more geographically distant, large-lake populations studied by Mort and Wolf (1985, 1986) were quite homogeneous with respect to gene frequencies. Why do the permanent *D. magna* populations exhibit so much more differentiation despite being subject to greater gene flow? A possible answer lies in the pattern of selection. The variance in the selection coefficients for the allelic variants in the *D. magna* populations is relatively high, whereas that for the large-lake populations of Mort and Wolf seems to be essentially zero (Lynch 1987). The observed differences in the geographic structure of gene frequencies in these two groups of populations are qualitatively consistent with the expectations based on differences in fluctuating selection intensity. In no case do we have evidence that selection is operating on the allelic markers themselves. More likely, as noted above, certain aspects of population structure, such as the incidence of sex, influence the associations between the markers and other loci upon which selection is acting. Presumably, the inevitable development of gametic-phase disequilibrium in response to clonal selection causes the variance in selection intensity on isozyme variants to be unusually high in *Daphnia* relative to sexual species.

Through the use of restriction-site mapping, we have extended our work on population-genetic structure in *D. pulex* to the mitochondrial DNA (Crease et al. 1990). Within populations, the average nucleotide diversity is approximately 0.002. In other words, random pairs of mitochondria sampled from an individual population differ at about 0.2% of their nucleotide sites. Estimates of this statistic for a diversity of other species range from 0.05% to 0.5% (Lynch and Crease 1990), so *D. pulex* is not particularly noteworthy in this regard. On the other hand, extension of the nucleotide diversity analysis to the between-population level has revealed this species to be among the most highly subdivided taxa ever observed at the level of mitochondrial DNA.

We measure population subdivision at the level of nucleotides by use of the index N_{st}, which has a zero-to-one scale as in the case of G_{st} (Lynch and Crease 1990). N_{st} is on the order of 0.2 for populations less than 1 km apart, and approaches an asymptote of approximately 0.7 for populations 100–1000 km apart. The latter value exceeds the differentiation for the entire New World population of *Drosophila melanogaster* (Lynch and Crease 1990).

The degree of geographic differentiation at the level of mitochondrial genes in *D. pulex* is approximately three times that of nuclear genes. Such a pattern is in rough accord with theoretical expectations. Since the effective population size for the mitochondrial genes is only one-quarter of that for genes in the nucleus, they are approximately four times as vulnerable to random genetic drift. Recall, however, that the between-population divergence of isozyme frequencies occurs more rapidly than expected on the basis of random genetic drift alone. This will almost certainly be true for the mitochondrial variants as well, since they will be subject to the same random processes that lead to disequilibrium with selected quantitative trait loci.

Recently, we extended this type of analysis to a multigene family—the 18s,28s ribosomal DNA (Crease and Lynch 1991). Among only ninety clonal isolates of *D. pulex*, thirty-seven distinct repeat types were identified, and depending on the population, individuals carried a minimum of two to four repeat types. The average nucleotide diversity for random repeats within populations was approximately 0.003, slightly greater than that observed for the mitochondrial genome. The degree of population subdivision was slightly less than the level observed for single-copy nuclear genes (isozymes). These results were somewhat surprising. It has been suggested that homogenizing forces such as gene conversion, unequal crossing-over, and replication slippage will tend to eliminate within-population diversity in multigene families while driving isolated populations to fixation for alternative copy types (Dover 1982; Ohta and Dover 1984). If that were true, then N_{st} for multigene families should be high compared to that for single-copy loci.

To evaluate the phylogenetic relationships between our three study species, we have computed Nei's (1972) genetic distances (D) for all combinations of populations. This statistic estimates the average number of allelic substitutions separating random pairs of genes sampled from two populations or species. For neutral markers, D is expected to increase linearly with time, provided the mechanisms underlying the molecular evolution remain constant throughout the phylogeny and the taxa under consideration remain isolated reproductively. When populations are connected via migration, D is expected to approach an equilibrium value defined by the balance between drift, mutation, and migration (Slatkin, chap. 1, this volume).

The fitted tree in figure 6.2 indicates that the distance of *D. obtusa* to *D. pulex–D. pulicaria* is approximately five times that between the latter two species. For the most part, the distance among populations of the same species is relatively small compared to that between taxa. However, there are some striking exceptions, leading to the impression that there is nearly a continuous distribution of genetic distances ranging from the within-species to the between-species level. For example, an Indiana population of *D. pulex* (PA in the figure) is quite distinct from five Illinois populations despite the fact that they are quite close geographically. An Oregon population (AMZ in the figure) that appears to be quite similar to *D. pulex* morphologically lies well outside of the *D. pulex-pulicaria* group.

Despite the fact that *D. pulex* and *D. obtusa* are separated by approxi-

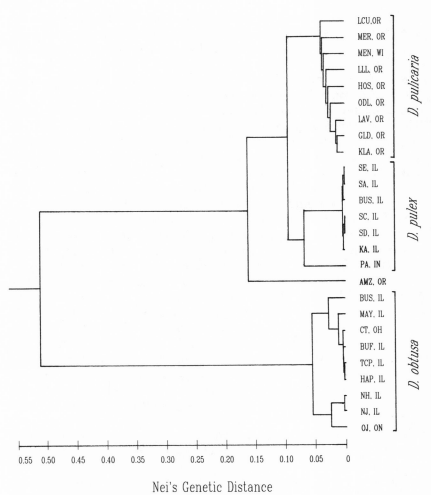

Nei's Genetic Distance

FIG. 6.2. Phylogenetic relationships between various populations of *D. obtusa*, *D. pulex*, and *D. pulicaria*. The total genetic distance between any two populations is equal to twice the distance to the common ancestor linking them. The phylogenetic tree was estimated by UPGMA. Each branch tip represents a population with the U.S. state abbreviation appearing after the comma (ON denotes Ontario, Canada).

mately 1.0 substitutions per isozyme locus, they can still be induced to hybridize in the laboratory (Agar 1920; Pojman, pers. obs.). However, the offspring of such matings do not appear to be capable of reproduction. On the other hand, *D. pulex* and *D. pulicaria* do appear to be cross-fertile. Extensive isozyme and mitochondrial DNA evidence indicates that such hybridizations have given rise to large numbers of obligately parthenogenetic clones throughout Canada and the midwestern United States (Crease et al. 1989, 1990; Hebert et al. 1989a).

Other *Daphnia* species are known to hybridize in nature—*D. hyalina, D. galeata,* and *D. cucullata* in Europe (Wolf 1987), *D. carinata* and *D. cephalata* in Australia (Benzie 1986b; Hebert 1985), and *D. galeata* and *D. rosea* in North America (Taylor and Hebert 1992). In each of these cases, the genetic distance between parental species is on the order of 0.3 to 0.4. Thus it appears that a genetic distance (based on isozymes) of 0.4 or so must be exceeded for reproductive isolation to be firmly established between *Daphnia* species. Such divergence may require a very long time. Based on empirical data in Nei (1987), time (in years) $\cong 5 \times 10^6 \times D$, suggesting that reproductive isolation in *Daphnia* species requires at least a few million years of isolation.

Life-History Evolution

Genetic diversity can be characterized in several ways at the level of quantitative traits. Within populations, the fraction of the total phenotypic variance attributable to genetic differences among clones is known as the broad-sense heritability,

$$H^2 = \frac{\sigma_{GW}^2}{\sigma_{GW}^2 + \sigma_E^2},$$

where σ_{GW}^2 is the total expressed genetic variance (the variance between clones) and σ_E^2 is the environmental component of variance for the trait (the variance within clones).

Our general approach to estimating H^2 in *Daphnia* has been to isolate random clones from a population and subsequently raise them in the laboratory under defined conditions. Prior to analysis, each clone is maintained as two or more sublines for two to three generations. This treatment insures that, in the final analysis, maternal effects and/or container effects do not contribute to the between-clone component of variance, as this would lead to overestimates of the genetic variation (Lynch and Ennis 1983; Lynch 1985). The experimental assay is performed on one or more replicates from each subline, with the entire collection of individuals being maintained in a randomized design in an environmental chamber. In the studies reported here, individuals were maintained at 20°C on a 12:12 light:dark cycle and fed a saturating level of a green alga. The genetic and environmental components of variance were obtained by equating the mean squares of an analysis of variance to their expectations. Further details of the laboratory and analytical methods can be found in Lynch (1985) and Lynch et al. (1989).

TABLE 6.1
Genetic diversity statistics for life-history characters

		L_{o2}	G_j	G_a	$C_{1,2}$	$C_{3,4}$	k_1	L_{k-1}
					Character[a,b]			
Daphnia pulex:								
H^2	PA[c]	0.44**	0.32**	0.36**	0.32**	0.42**	0.04	0.33**
	KA	0.50**	0.52**	0.59**	0.41**	0.54**	0.43**	0.40**
Daphnia pulicaria:								
H^2	HO	0.25	0.01	−0.10	0.08	0.14	0.53**	−0.16
	OD	0.31	0.09	0.55**	0.00	−0.01	0.03	0.18
	KL	0.29	0.44*	0.13	0.69**	0.71**	−0.12	0.30**
Daphnia "amazon":[d]								
H^2	AMZ	0.32**	0.38**	0.79**	0.22	0.34**	0.38**	0.20
Daphnia obtusa:								
H^2		0.48**	0.22*	0.05	0.31**	0.37**	0.30	0.17*
Q_{ST}		0.31**	0.17**	0.41**	0.19*	0.18**	0.20*	0.58**

[a] *Key to characters*: L_{o2} is the mean length (mm) of offspring produced in the second clutch; G_j = $\ln(L_k/L_0)/t$ is the juvenile growth rate, with L_0 and L_k being the lengths at birth and at the time of carrying the first clutch, and t the time between these points; $G_a = \ln(L_{k+3}/L_k)/t$ is the adult growth rate, with L_{k+3} being the length in the fourth adult instar; $C_{1,2}$ refers to the number of offspring released in the first and second clutches (statistics were computed separately for each trait and then pooled); $C_{3,4}$ is defined similarly for the third and fourth clutches; k_1 is the age at first release of live progeny; and L_{k-1} is the length in the instar prior to the first appearance of a clutch.

[b] *Key to significance levels*: ** denotes $P \leq 0.01$, * denotes $0.01 < P \leq 0.05$.

[c] The different study populations are PA (Portland Arch, Indiana), KA (Kickapond, Illinois), HO (Hosmer Lake, Oregon), OD (Odell Lake, Oregon), KL (Klamath Lake, Oregon), and AMZ (Amazon Park, Oregon).

[d] The Amazon population is electrophoretically distinct enough from the other species in this survey that it probably is a unique, but undescribed, species. We therefore refer to it as *D. "amazon"* throughout.

The heritability estimates given in table 6.1 for *D. pulex*, *D. "amazon,"* and *D. obtusa* were obtained from collections of clones taken from temporary ponds early in the spring, shortly after the hatching of resting eggs. (The heritability estimates for *D. obtusa* are pooled results from a simultaneous analysis of eight populations.) The heritabilities of most size, growth, and reproductive traits in these populations are highly significant, ranging from 0.2 to 0.8. Qualitatively similar results were obtained earlier, with slightly different methods, for another temporary pond population of *D. pulex* (Lynch 1984b).

We have performed another assay of this sort with animals from a *D. pulex* population known to be reproducing by obligate parthenogenesis. Although the experiment was comparable in power to those described above, none of the twenty life-history traits examined exhibited significant heritability (Lynch et al. 1989). Similarly, when assays were performed on collections of clones taken from a cyclically parthenogenetic population toward the end of the growing season, the heritabilities were found consistently to be much lower than those found several weeks earlier in the same pond (Lynch 1984b). Both observations are in agreement with the idea that prolonged clonal selection is very effective at eliminating most of the genetic variation for life-history traits from a population. This hypothesis is supported further by results from laboratory experiments involving electrophoretically marked populations of clones (Spitze 1991).

The *D. pulicaria* populations that we have examined occupy large permanent lakes in the Oregon Cascades, and it is likely that individuals overwinter in the water column. The analyses were again comparable in size and design to those noted above, yet for two of the populations surveyed (Hosmer and Odell), only a single character exhibits significant heritability (table 6.1). It seems likely that prolonged periods of clonal selection in the absence of significant recruitment from resting eggs may serve to keep quantitative-genetic variation low in these populations, similar to the situation in populations reproducing by obligate parthenogenesis. Both populations exhibited significant deviations from Hardy-Weinberg and gametic-phase equilibria for electrophoretic markers (Lynch et al., in prep.), consistent with the expectation for populations that have not been homogenized recently by sexual reproduction. Interestingly, a third *D. pulicaria* population, from Klamath Lake, harbored substantial genetic variation for life-history traits (table 6.1). Unlike the Hosmer and Odell populations, this population was in Hardy-Weinberg equilibrium, suggesting that it may have recently experienced a phase of sexual reproduction.

At the species level, the total genetic variation can be partitioned into within- and between-population components. The fraction of the total genetic variation attributable to differences among populations, which we denote as Q_{ST}, provides a useful phenotypic analog of G_{ST} and N_{ST}. Q_{ST} takes a slightly different form than our measures of gene and nucleotide diversity. For quantitative characters with an additive genetic basis, Wright (1951) showed that the mean additive genetic variance within populations is $\overline{\sigma_{Gw}^2} = (1 - Q_{ST})\sigma_G^2$, where σ_G^2 is the additive genetic variance that would exist if all populations were joined by random mating into one panmictic unit. The expected between-population variance is $\sigma_{GB}^2 = 2Q_{ST}\sigma_G^2$. It follows that

$$Q_{ST} = \frac{\sigma_{GB}^2}{2\overline{\sigma_{Gw}^2} + \sigma_{GB}^2}.$$

For a character with an additive-genetic basis in a diploid species, Q_{ST} has exactly the same expected value that G_{ST} would have if the latter were computed on the basis of the allele frequencies for the quantitative-trait loci. The two in front of $\overline{\sigma^2_{Gw}}$ is due to the fact that Q_{ST} is based on a comparison of genotypes, whereas G_{ST} and N_{ST} are based on comparisons of genes; the between-population variance for a quantitative trait is magnified due to the statistical association of identical genes within individuals in subdivided populations.

Since the *D. obtusa* study was based on eight populations, it was possible to estimate Q_{ST} for life-history traits. Population differentiation was significant for all of the characters, even for the ones with negligible variation within populations (table 6.1). Remarkably, the average value of Q_{ST}, 0.29(0.06), is the same as the value of G_{ST} obtained for this species with isozymes. Since, as noted above, the population divergence of isozyme frequencies in *Daphnia* almost certainly is inflated above the neutral expectation by fluctuating selection, these results suggest that similar phenomena may be playing a role in diversifying the mean phenotypes of different populations. This is supported by recent results of a clonal analysis of populations of *D. pulicaria* and *D. galeata* in several Michigan lakes. Using the data of Leibold and Tessier (1991), the estimates of Q_{ST} are 0.46(0.05) and 0.61(0.24), respectively. Again, compared to the estimates of G_{ST} reported above for broader geographic regions (fig. 6.1), this amount of population subdivision is quite high. Leibold and Tessier (1991) provide compelling evidence that the divergence is due to the local adaptation of populations to selection by vertebrate predators.

We have attempted to put the observed divergence among species in a population-genetic context by computing for each trait the divergence statistic of Lynch (1990),

$$\Delta = \frac{\sigma^2_{GB}}{T\sigma^2_E},$$

where T is the total number of generations of divergence between a pair of species. Δ is a useful measure because of its relationship to the neutral expectation. In the absence of natural selection, the variance among the mean phenotypes of isolated taxa is expected to increase at the rate $2\sigma^2_m$ per generation, where σ^2_m is the rate at which new variation enters a population per generation via mutation (Lynch, chap. 5, this volume). Estimates of σ^2_m/σ^2_E, where as above σ^2_E is the environmental variance for the trait, range from 10^{-4} to 10^{-2} for a wide range of characters and species, including *Daphnia* (Lynch 1985, 1988). Thus, observed values of Δ below 2×10^{-4} are consistent with the hypothesis that natural selection

TABLE 6.2
Mean phenotypes for the species, and the divergence statistic for
D. obtusa–D. pulex/pulicaria

	Character[a]						
	L_{o2}	G_j	G_a	C_1	C_3	k_1	L_{k-1}
Mean phenotypes:							
D. pulex	0.628	0.242	0.038	14.5	35.6	6.78	1.57
	(0.025)	(0.006)	(0.002)	(2.5)	(1.2)	(0.08)	(0.01)
D. pulicaria	0.643	0.188	0.039	8.9	19.0	7.39	1.37
	(0.020)	(0.026)	(0.003)	(1.8)	(4.1)	(0.39)	(0.14)
D. "amazon"	0.549	0.135	0.030	7.7	18.3	10.04	1.40
	(0.007)	(0.004)	(0.002)	(0.5)	(1.3)	(0.26)	(0.02)
D. obtusa	0.599	0.228	0.047	13.2	31.7	6.64	1.40
	(0.008)	(0.003)	(0.001)	(0.4)	(1.7)	(0.08)	(0.03)
$\Delta(\times 10^{-8})$	3.2	0.0^b	0.5	0.0	0.0	0.2	0.0

[a] See table 6.1 for character descriptions.

[b] All Δ values reported as 0.0 were actually less than zero; the variance between observed means was less than the sampling variance of the mean.

has prevented species' mean phenotypes from diverging as fast as would be likely if random drift and mutation were the sole evolutionary forces.

Since *D. pulex* and *D. pulicaria* are not completely reproductively isolated, and the taxonomic status of *Daphnia "amazon"* is undetermined, we will simply consider the split between *D. obtusa* and *D. pulex/pulicaria* (averaging the mean phenotypes of the latter two species). Assuming that the populations average about four generations per year, then the genetic distance estimates given in figure 6.2 imply that the *D. obtusa–D. pulex/pulicaria* split occurred roughly $T = 2 \times 10^7$ generations in the past. The estimates of Δ given in table 6.2 were obtained by using half the squared difference between means, minus the sampling variance of the means, as an estimate of σ^2_{GB}. For each character, the environmental variance was taken to be the average value of the within-clone variances taken over populations and species.

Averaging over all seven characters, the mean value of Δ is $5.5(4.4) \times 10^{-9}$, which is roughly five orders of magnitude below the minimum neutral expectation. Such an extremely low rate of evolution strongly suggests that stabilizing selection plays a major role in preventing the diversification of size, growth, and reproductive characters among *Daphnia* species. Thus, while the high levels of genetic variation observed within and between populations of the same species indicate that these characters can diverge rapidly in response to local selection pressures, the

TABLE 6.3
Genetic correlations with adult size (L_{k-1})

	Character[a,b]					
	L_{o2}	G_j	G_a	$C_{1,2}$	$C_{3,4}$	k_1
Daphnia pulex:						
PA	0.60*	0.62*	−0.65*	0.33	0.18	−0.90
KA	0.57*	0.58**	−0.04	0.37*	0.71**	0.01
Daphnia pulicaria:						
KL	0.29	0.44*	0.13	0.69**	0.71**	−0.12
Daphnia "amazon":						
AMZ	0.14	0.26	0.03	0.48**	0.64**	−0.56
Daphnia obtusa:						
Within	1.00**	0.43	1.00**	0.38	0.75*	−0.42
Between	0.98**	0.75**	0.45	0.84**	0.91**	0.24

[a] *Key to significance levels*: ** denotes $P \leq 0.01$, * denotes $0.01 < P \leq 0.05$.
[b] See table 6.1 for a description of the characters.

amount of additional divergence that occurs after speciation is relatively small.

Body size has long been regarded as a fundamental determinant of the ecological and evolutionary properties of planktonic cladocerans (Brooks and Dodson 1965; Lynch 1980). To evaluate the degree to which natural selection on body size is expected to influence the evolution of other life-history characters, we have computed the genetic correlations between length at maturity (L_{k-1}) and the other characters in table 6.1. The genetic covariances between characters are estimated in a manner analogous to the univariate procedure for variance component estimation—by equating the mean cross-products of a multivariate analysis of variance to their expectations. Significant correlations at the genetic level imply that the traits are not free to evolve independently, either because of pleiotropic effects of segregating alleles or because of gametic-phase disequilibrium between variants at different loci.

For most of the populations we have observed, size at maturity is positively genetically correlated with size at birth, juvenile growth rate, and clutch sizes, and negatively genetically correlated with the age at first reproduction (table 6.3). On the other hand, the genetic correlation between adult size and adult growth rate appears to be evolutionarily labile; it is significantly positive in *D. obtusa* and significantly negative in *D. pulex* from Portland Arch. For *D. obtusa*, it was also possible to compute genetic correlations at the between-population level. These were all qualitatively consistent with the within-population estimates (table 6.3).

These results bear out the conclusion of a more extensive study of the genetic covariances of several life-history traits in *D. pulex* (Spitze et al. 1991)—contrary to the usual expectation (Travis, chap. 9, this volume), genetic trade-offs between life-history characters appear to be rare in *Daphnia*.

Discussion

A pattern that emerges from our studies of quantitative variation in *Daphnia* is that, for populations that go through an annual bout of sexual reproduction, pronounced levels of genetic variance for life-history characters early in the growing season are followed by nearly undetectable levels only a few weeks later. Such cycles of genetic variation can be repeated on an annual basis (Lynch 1984b). These observations are concordant with the predictions of a quantitative-genetic model for phenotypic evolution in cyclical parthenogens under stabilizing selection (Lynch and Gabriel 1983). Clonal selection effectively eliminates deviant genotypes, thereby rapidly reducing the level of expressed genetic variance. But because individual genotypes are discriminated solely on the basis of their phenotypic properties, irrespective of the underlying genotype, hidden genetic variance builds up throughout the period of clonal selection. For polygenic characters, the same phenotype can be obtained with many different genetic constitutions. Thus, in principle, it is possible for a population to exhibit no variation at the phenotypic level despite the existence of substantial variation at the molecular level. Our empirical observation of relatively stable genotype frequencies at the isozyme level (close to Hardy-Weinberg expectations) during a period of rapid loss of quantitative genetic variation confirms the idea that populations can consist of large numbers of ecologically equivalent clones at least for moderate periods of time (Lynch 1984a,b).

So long as a population is reproducing parthenogenetically, the pool of hidden genetic variance is evolutionarily inert, despite the fact that it continues to grow. However, a single bout of sex is sufficient to convert a large fraction of the hidden genetic variance into the expressed form (Lynch and Gabriel 1983). Thus, populations that abstain from sex for very long periods of time are expected to achieve much higher levels of expressed genetic variance, and hence higher short-term evolutionary potential, following the event than can ever be obtained in a purely sexual population.

If the prevailing form of selection is either diversifying or directional with a sufficiently concave fitness function, the results outlined above can be altered. Under these conditions, clonal selection will actually favor the coupling of genes of like effects. When segregation and recombination remove such positive associations, the variance among progeny clones can then be less than that among the parents. Such changes have been documented recently in the D. *"amazon"* population (Lynch and Deng, in prep.).

Under certain conditions, recruitment of sexually produced progeny can induce a change in the mean of a quantitative trait as well as in the variance, leading to genetic slippage from the end of one year to the beginning of the next. This can happen if a portion of the flush of genetic variance at the outset of each growing season is a consequence of the hatching of resting eggs produced over a period of several years. There is good evidence that the sediments in permanent lakes do contain a cladoceran "seed bank" (Herzig 1985; Carvalho and Wolf 1989). Thus, if ecological conditions are such that different mean phenotypes are favored in different years, while enhancing the amount of variation upon which selection can act, recruitment from multiple cohorts of resting eggs could significantly erode the selective progress made in the preceding year.

A second issue that remains to be resolved at both the empirical and theoretical levels is the extent to which nonadditive gene action contributes to the evolutionary dynamics of phenotypic means and variances. As noted above, clonal selection operates on the total genotypic properties of individuals, whereas selection in a sexual population advances alleles primarily on the basis of their additive effects. If significant nonadditive gene combinations are favored by selection, recombinational breakdown will occur upon sexual reproduction. Thus, depending upon the mode of gene action, long-term clonal selection can facilitate the evolution of coadapted gene complexes, only to be followed by a sort of outbreeding depression upon sexual reproduction. The fact that *Daphnia* exhibit inbreeding depression (Banta 1939; Innes 1989), which is not possible with purely additive gene action, implies that these issues warrant further exploration. Recently in the D. *"amazon"* population, we observed genetic slippage in the means for life-history traits averaging about 10% of the phenotypic standard deviations (Lynch and Deng, in prep.). Future work is needed to evaluate the extent to which the magnitude of genetic slippage builds up with the length of the asexual phase and to determine the degree to which genetic slippage erodes the response to directional selection during the clonal phase.

So far, our results indicate that the genetic architectures (heritabilities and genetic correlations) of different populations of the same species are qualitatively similar (Spitze et al. 1991). This conclusion is upheld for the most part when the genetic correlations of different species are compared, suggesting that the microevolutionary responses of different species to the same selection pressures will be qualitatively similar as well. With the possible exception of adult growth rate, the entire constellation of characters that we have examined appears to evolve in a highly coordinated fashion. It remains to be seen whether these parallel patterns are a consequence of similar pleiotropic constraints throughout the entire assemblage or of the overriding influence of selection favoring specific allometric relationships.

A common observation is that whereas microevolutionary change among different populations of the same species can proceed quite rapidly, there is a progressive slowdown in the rate of phenotypic divergence as an established phylogenetic group ages (Lynch 1990). Our results are certainly consistent with this pattern. Individual populations harbor substantial genetic variation for quantitative traits, and this permits a rather high degree of divergence among isolated populations. However, the evolutionary changes in life-history features among species appear to be within the evolutionary potential of individual species. Reproductive isolation has done essentially nothing to facilitate the divergence of life-history characters among the species we have studied.

Acknowledgments

This work has been supported by NSF grant BSR 89-11038 and PHS grant R01 GM36827-01 to Michael Lynch, and a Summer Award in Natural Sciences from the Research Council of the University of Miami to Ken Spitze. The work could not have been completed without the help of a large number of people. We are especially grateful to M. Berigan, T. Crease, W. Gabriel, R. Gibson, B. Hecht, G. Henderson, T. Leatham, B. Monson, C. Smythe, A. Toline, and L. Weider. We thank E. Martins for helpful comments.

References

Agar, W. E. 1920. The genetics of a *Daphnia* hybrid during parthenogenesis. *J. Genet.* 10: 303–330.

Banta, A. M. 1939. Studies on the physiology, genetics, and evolution of some Cladocera. Paper No. 39, Dept. of Genetics, Carnegie Institution of Washington, D.C.

Benzie, J.A.H. 1986a. The ecological genetics of freshwater zooplankton in Australia. In *Limnology in Australia*, ed. P. DeDeckker and W. D. Williams, pp. 175–191. W. J. Junk, Dordrecht.

Benzie, J.A.H. 1986b. Phylogenetic relationships within the genus *Daphnia* (Cladocera: Daphniidae) in Australia, determined by electrophoretically detectable protein variation. *Aust. J. Mar. Freshw. Res.* 37: 251–260.

Brooks, J. L., and S. I. Dodson. 1965. Predation, body size, and composition of plankton. *Science* 150: 28–35.

Carvalho, G. R., and H. G. Wolf. 1989. Resting eggs of lake-*Daphnia*. I. Distribution, abundance and hatching of eggs collected from various depths in lake sediments. *Freshw. Biol.* 22: 459–470.

Crease, T. J., and M. Lynch. 1991. Ribosomal DNA variation in *Daphnia pulex*. *Mol. Biol. Evol.* 8: 620–640.

Crease, T. J., M. Lynch, and K. Spitze. 1990. Hierarchical analysis of population genetic variation in mitochondrial and nuclear genes of *Daphnia pulex*. *Mol. Biol. Evol.* 7: 444–458.

Crease, T. J., D. J. Stanton, and P.D.N. Hebert. 1989. Polyphyletic origin of asexuality in *Daphnia pulex*. II. Mitochondrial DNA variation. *Evolution* 43: 1016–1026.

Dover, G. A. 1982. Molecular drive: A cohesive mode of species evolution. *Nature* 299:111–117.

Hebert, P.D.N. 1974a. Enzyme variability in natural populations of *Daphnia magna*. II. Genotypic frequencies in permanent populations. *Genetics* 77: 323–334.

Hebert, P.D.N. 1974b. Enzyme variability in natural populations of *Daphnia magna*. II. Genotypic frequencies in intermittent populations. *Genetics* 77: 335–344.

Hebert, P.D.N. 1981. Obligate asexuality in *Daphnia*. *Amer. Nat.* 117: 784–789.

Hebert, P.D.N. 1985. Interspecific hybridization between cyclic parthenogens. *Evolution* 39: 216–219.

Hebert, P.D.N., R. D. Ward, and L. J. Weider. 1988. Clonal-diversity patterns and breeding-system variation in *Daphnia pulex*, an asexual-sexual complex. *Evolution* 42: 147–159.

Hebert, P.D.N., M. J. Beaton, and S. S. Schwartz. 1989a. Polyphyletic origin of asexuality in *Daphnia pulex*. I. Breeding system variation and levels of clonal diversity. *Evolution* 43: 1004–1015.

Hebert, P.D.N., S. S. Schwartz, and J. Hrbacek. 1989b. Patterns of genotypic diversity in Czechoslovakian *Daphnia*. *Heredity* 62: 207–216.

Hebert, P.D.N., S. S. Schwartz, and L. J. Weider. 1989c. Geographical patterns in genetic diversity and parthenogenesis within the *Daphnia pulex* group from the southern United States. *Amer. Midl. Nat.* 122: 59–65.

Herzig, A. 1985. Resting eggs—a significant stage in the life cycle of the crustaceans *Leptodora kindti* and *Bythotrephes longimanus*. *Verh. Internat. Verein. Limnol.* 22: 3088–3098.

Innes, D. J. 1989. Genetics of *Daphnia obtusa*: Genetic load and linkage analysis in a cyclical parthenogen. *J. Hered.* 80: 6–10.

Kimura, M. 1954. Process leading to quasi-fixation of genes in natural populations due to random fluctuations of selection intensities. *Genetics* 39: 280–295.

Kimura, M. 1983. *The Neutral Theory of Molecular Evolution*. Cambridge University Press, Cambridge, U.K.

Korpelainen, H. 1984. Genic differentiation of *Daphnia magna* populations. *Hereditas* 101: 209–216.

Korpelainen, H. 1986a. Genetic variation and evolutionary relationships within four species of *Daphnia* (Crustacea: Cladocera). *Hereditas* 105: 245–254.

Korpelainen, H. 1986b. Ecological genetics of the cyclic parthenogens, *Daphnia longispina* and *Daphnia pulex*. *Hereditas* 105: 7–16.

Korpelainen, H. 1986c. Temporal changes in the genetic structure of *Daphnia magna* populations. *Heredity* 57: 5–14.

Leibold, M., and A. J. Tessier. 1991. Contrasting patterns of body size for *Daphnia* species that segregate by habitat. *Oecologia* 86: 342–348.

Lynch, M. 1980. The evolution of cladoceran life histories. *Quart. Rev. Biol.* 55: 23–42.

Lynch, M. 1983. The ecological genetics of *Daphnia pulex*. *Evolution* 37: 358–374.

Lynch, M. 1984a. The genetic structure of a cyclical parthenogen. *Evolution* 38: 186–203.

Lynch, M. 1984b. The limits to life history evolution in *Daphnia*. *Evolution* 38: 465–482.

Lynch, M. 1985. Spontaneous mutations for life-history characters in an obligate parthenogen. *Evolution* 38: 804–818.

Lynch, M. 1987. The consequences of fluctuating selection for isozyme polymorphisms in *Daphnia*. *Genetics* 115: 657–669.

Lynch, M. 1988. The rate of polygenic mutation. *Genet. Res.* 51: 137–148.

Lynch, M. 1990. The rate of morphological evolution in mammals from the standpoint of the neutral expectation. *Amer. Nat.* 136: 727–741.

Lynch, M., and T. J. Crease. 1990. The analysis of population survey data on DNA sequence variation. *Mol. Biol. Evol.* 7: 377–394.

Lynch, M., and R. Ennis. 1983. Resource availability, maternal effects, and longevity. *Exper. Geront.* 18: 147–165.

Lynch, M., and W. Gabriel. 1983. Phenotypic evolution and parthenogenesis. *Amer. Nat.* 122: 745–764.

Lynch, M., K. Spitze, and T. Crease. 1989. The distribution of life-history variation in the *Daphnia pulex* complex. *Evolution* 43: 1724–1736.

Mort, M. A., and H. G. Wolf. 1985. Enzyme variability in large-lake *Daphnia* populations. *Heredity* 55: 27–36.

Mort, M. A., and H. G. Wolf. 1986. The genetic structure of large-lake *Daphnia* populations. *Evolution* 40: 756–766.

Nei, M. 1972. Genetic distance between populations. *Amer. Nat.* 106: 283–292.

Nei, M. 1987. *Molecular Evolutionary Genetics*. Columbia University Press, New York.

Ohta, T., and G. A. Dover. 1984. The cohesive population genetics of molecular drive. *Genetics* 108: 501–521.

Selander, R. K., A. G. Clark, and T. S. Whittam. 1991. *Evolution at the Molecular Level*. Sinauer, Sunderland, Mass.

Spitze, K. 1991. *Chaoborus* predation and life-history evolution in *Daphnia pulex*: Temporal pattern of population diversity, fitness, and mean life history. *Evolution* 45: 82–92.

Spitze, K., J. Burnson, and M. Lynch. 1991. The covariance structure of life-history characters in *Daphnia pulex*. *Evolution* 45: 1081–1090.

Taylor, D., and P.D.N. Hebert. 1992. *Daphnia galeata mendotae* as a cryptic species complex with interspecific hybrids. *Limnol. Oceanogr.* 37: 658–665.

Weider, L. J. 1985. Spatial and temporal genetic heterogeneity in a natural *Daphnia* population. *J. Plankton Res.* 7: 101–123.

Weir, B. S. 1990. *Genetic Data Analysis*. Sinauer, Inc. Sunderland, Mass.

Wolf, H. G. 1987. Interspecific hybridization between *Daphnia hyalina*, *D. galeata*, and *D. cucullata* and seasonal abundances of these species and their hybrids. *Hydrobiol.* 145: 213–217.

Wright, S. 1948. On the roles of directed and random changes in gene frequency in the genetics of populations. *Evolution* 2: 279–294.

Wright, S. 1951. The genetic structure of populations. *Annals of Eugenics* 15: 323–354.

Young, J.P.W. 1979a. Enzyme polymorphism and cyclic parthenogenesis in *Daphnia magna*. I. Selection and clonal diversity. *Genetics* 92: 953–970.

Young, J.P.W. 1979b. Enzyme polymorphism and cyclic parthenogenesis in *Daphnia magna*. I. Heterosis following sexual reproduction. *Genetics* 92: 971–982.

7

The Interplay of
Numerical and Gene-Frequency Dynamics
in Host-Pathogen Systems

Introduction

We know very little about how pathogens affect natural plant populations, but we have a vast body of literature on how pathogens affect crop
plants. The neglect of pathogens by plant ecologists is curious, but it is
consistent with the historical trend where, in the past, primary attention
was given to examining the role of abiotic factors in plant distribution.
Only within the past two decades has explicit focus been placed on biotic
factors such as plant-plant competition and plant-insect interactions. The
current surge of interest in plant-pathogen interactions (Burdon 1987;
Fritz and Simms 1992) is therefore an extension of the realization that
biotic forces, while perhaps of secondary consideration with regard to
explaining differences among the major biomes and vegetation types,
may be of primary importance in determining species abundance and diversity within those biomes. Unfortunately, the knowledge acquired in an
agricultural setting has been gained in the context of a research tradition
based on the applied goals of crop production and cannot be easily ex-
trapolated to provide an understanding of natural systems. For example,
agriculture has been concerned with disease control, with its emphasis on
disease symptoms rather than on host and pathogen fitnesses; it has focused on situations where crops are planted at set densities and where the
major dynamic processes are the epidemic spread of pathogens through
preexisting monospecific, genetically uniform stands; and because agriculture is concerned with crop yield, a group property, there has been
naive group-selection thinking about evolutionary outcomes (see Antonovics and Alexander 1989, for a discussion of these issues). In addition, at least in the United States, the lack of involvement of agricultural
funding agencies (e.g., United States Department of Agriculture) in basic
research combined with the concentration of plant pathologists in agricultural departments has severely aggravated the disjunction between the
pure and applied aspects of the subject.

The applied study of plant pathogens has nevertheless provided two
obvious but major generalizations that can serve as a starting point for

extrapolations to natural systems. The first general observation is that disease can have an enormous impact on the reproduction and survival of individual plants. The second general observation is that individual genotypes of both host and pathogen may differ greatly in resistance and virulence (Day 1974; Vanderplank 1984; Wolfe and Caten 1987). Given these two observations, we can expect host-pathogen dynamics to be driven by the interaction of strong ecological and genetic forces. We can expect dramatic impacts on survival and reproduction, unforgiving selection pressures, and consequently large changes in population size and gene frequency. In this and the following chapter, I explore these expectations further.

Any host-pathogen system can vary in many ways: the disease may or may not cause changes in fecundity or survival, individuals may or may not recover from the disease, and there may or may not be induced resistance (i.e., the equivalent of an "immune class"). Because the possibilities are many, I focus here on models based on a real-world example that I have been studying, namely the anther-smut disease of white campion, *Silene alba* (= *S. latifolia*) caused by the fungus *Ustilago violacea*. I describe the biology of this system in more detail in the next chapter, but the salient features are (1) the disease sterilizes the host because both male and female plants develop anthers bearing spores and the ovary aborts in female flowers, (2) the disease is systemic such that the host usually does not recover, and (3) transmission of the disease is by pollinators. Because pollinators adjust their flight distances to compensate for plant density, it is likely that the probability of a healthy individual becoming diseased is a function of the frequency (and not absolute density) of diseased individuals in the population. In experimental arrays of diseased and healthy plants, spore deposition on healthy plants increased with increasing frequency of diseased plants but not with increasing population density (Antonovics and Alexander 1992). Some spore dispersal also occurs by spores falling from the parent plant, and seedlings placed in close vicinity to the parent can become infected (Alexander 1990). However, there is no evidence that the disease is transmitted through the seeds, even when those seeds are produced from partially diseased plants (Baker 1947). The pathogen may decrease plant longevity to some degree (Alexander and Antonovics 1988; Alexander 1989; Thrall, pers. comm.), but plants with the disease appear more or less normal vegetatively. The host has been shown to have genetic variation in resistance and susceptibility, whereas the fungus appears to be quite uniform genetically (Alexander, Antonovics, and Kelly 1993). There also appears to be some resistance cost; male plants that are more resistant have fewer flowers and flower later in the season (Alexander 1989).

Fortunately, these aspects of the biology of the *Silene-Ustilago* system can be encapsulated into fairly simple but general models. Thus throughout we will assume that there is no reproduction of diseased hosts, that hosts do not recover from the disease, and that there is no disease-induced mortality or immunity. With regard to the genetics, we will assume that the host is genetically variable for resistance but the pathogen is genetically uniform. We will also confine our attention to single isolated populations, realizing (as outlined in the following chapter) that this may inadequately describe the disease dynamics on a more regional basis.

Frequency, Density, and the Transmission Process

Population geneticists have traditionally focused on the dynamics of gene frequencies, whereas population ecologists have focused on the dynamics of numerical change. The numerical dynamics of the *Silene-Ustilago* system can be represented by simple difference equations:

$$X_{t+1} = X_t (1 + b - d) - X_t P_t \qquad (7.1a)$$
$$Y_{t+1} = Y_t (1 - d) + X_t P_t \qquad (7.1b)$$

where X, Y = number of healthy and diseased individuals

b = birth rate of healthy individuals

d = death rate of diseased and healthy individuals

P = probability a healthy individual becomes diseased

t = time interval.

For simplicity, we assume here that birth, death, and infection processes occur simultaneously, but we could easily modify the equations to represent situations where, for example, plants that become diseased early in the time interval do not subsequently reproduce.

The transmission parameter, P, can take two forms depending on whether one views disease transmission as a function of the absolute density of diseased plants,

$$P_t = \beta Y_t, \qquad (7.2a)$$

or the frequency of diseased plants,

$$P_t = \beta Y_t / N_t. \qquad (7.2b)$$

The density-dependent case would represent aerial transmission of spores where the per capita likelihood of a healthy individual becoming diseased increases as the density of diseased individuals increases. The frequency-dependent case would represent pollinator transmission of spores, where the pollinator increases its flight distance to compensate for increases in plant spacing: this is the predominant mode of transmission of anther-smut in *Silene*. This transmission mode is likely to be a general characteristic of vector-borne and venereal diseases. A number of previous studies have shown that the transmission mode can have a large effect on the population dynamics (Getz and Pickering 1983; Thrall, Antonovics, and Hall 1992).

Extension of these cases to include genetic variation in host and pathogen is straightforward. This can be best illustrated by assuming the host is haploid and has two alleles at one locus:

$$X1_{t+1} = X1_t(1 + r1 - P1_t) \tag{7.3a}$$

$$X2_{t+1} = X2_t(1 + r2 - P2_t) \tag{7.3b}$$

$$Y1_{t+1} = Y1_t(1 - d) + X1_t P1_t \tag{7.3c}$$

$$Y2_{t+1} = Y2_t(1 - d) + X2_t P2_t, \tag{7.3d}$$

where $r = b - d$, and labels 1 and 2 refer to two different haploid genotypes of the host.

We can easily reformulate these equations in terms of frequency and not absolute number of the different types (setting $x, y = X/N, Y/N$):

$$Tx1_{t+1} = x1_t(1 + r1 - \beta1 y_t) \tag{7.4a}$$

$$Tx2_{t+1} = x2_t(1 + r2 - \beta2 y_t) \tag{7.4b}$$

$$Ty1_{t+1} = y1_t((1 - d) + \beta1 x1_t) \tag{7.4c}$$

$$Ty2_{t+1} = x2_t((1 - d) + \beta2 x2_t), \tag{7.4d}$$

where $T = 1 + r1 \, x1_t + r2 \, x2_t - d(y1_t + y2_t)$.

At least superficially, it may seem that by restricting ourselves to only gene-frequency dynamics, we are only "standardizing" frequencies by total population size, and therefore simplifying the models. However, it can be seen (eqs. 7.4a–d) that the distinction between the density- and frequency-dependent transmission modes is lost. Indeed, models that restrict themselves to consideration of only gene-frequency dynamics implicitly assume that the transmission mode is frequency-dependent. Such models can no longer directly address the dynamics of systems with density-dependent transmission. In using models based on gene frequency, we are also making implicit assumptions about population size and/or its regulation. Formally, these assumptions are either that the population size is constant (it is set to unity at each generation), or that it is changing in an unregulated manner (there is no density-dependent population regulation), or that population regulation acts proportionately on the healthy and diseased classes. An example of the latter may be where density dependence acts equally on, say, overwinter mortality (time interval $t + 1, t + \frac{1}{2}$) of both diseased and healthy individuals, but the remaining dynamics is determined in the summer period (time interval $t + \frac{1}{2}, t$) by processes described using equations of the form (7.1) to (7.4) above. Therefore, models that do not incorporate numerical dynamics assume that density-dependent regulation is either not occurring or is proportionate among the types, and that the disease transmission process is frequency-dependent.

Population Regulation and Coexistence

It is commonly appreciated that the stability and dynamics of host-pathogen and predator-prey systems can be greatly affected by the presence of extrinsic factors limiting host and pathogen abundance (Hassell 1978; Anderson and May 1978). As mentioned above, for these extrinsic factors to have an impact, they must influence the rate of change of healthy and diseased individuals differentially. This can come about in two ways. First, the density-dependent processes may act equally on the birth and death rates of healthy and diseased individuals, but they may not affect the disease transmission dynamics. This introduces a nonproportionality in the rate of growth of the healthy and diseased classes. Second, the degree to which density alters the birth and death rates may differ among healthy and diseased individuals. Most generally we might expect diseased individuals to be more susceptible to the impact of density. A number of studies have shown that diseased individuals are competitively inferior to healthy individuals (Reestman 1946, cited in De Wit 1960; Burdon and Chilvers 1977) but I know of no explicit studies comparing the impact of density on diseased and healthy individuals. A number of other examples are known where the presence of a disease protects the host against predators and/or improves their competitive abilities (Clay 1991). If there is age or stage specificity in the probability of becoming diseased, but the impact of density occurs at an earlier or different stage, then healthy and diseased individuals may also have different density-dependent responses. This is likely to be the case in the *Silene-Ustilago* system as well as in other venereal disease systems where disease transmission is in the adult reproductive phase, but where density has the greatest impact on the juvenile phase.

The importance of differential density dependence is well illustrated in models of the type we have outlined above (eqs. 7.1, 7.2). Without differential density dependence, and given the assumptions of no host recovery and no reproduction of diseased individuals, density-dependent disease transmission predicts neutral stable limit cycles. With linear density dependence acting on reproductive rates or death rates, there is an oscillatory approach to equilibrium (Antonovics 1992). Because the fungus sterilizes the host and converts it morphologically entirely for its own reproduction, it is in a sense a predator or parasitoid, and the models developed above are similar in form to those developed for predator-prey or host-parasitoid systems (Hassell 1978). In frequency-dependent transmission systems (e.g., venereal diseases), equilibrium coexistence of host and pathogen is not possible unless one invokes some external density-dependent factor limiting host and pathogen mortality or reproduction (Getz and Pickering 1983). However, if there is no host recovery and no reproduction of diseased individuals, as we assume in our *Silene-Ustilago* system, then host-pathogen coexistence further requires that the degree of density dependence be different in the healthy and diseased class (Thrall, Antonovics, and Hall 1992). In the case of sterilizing venereal diseases such as the anther-smut of *Silene*, unequal density-dependent effects greatly expand the range of parameters (disease transmission rate, death rate, and reproductive rate) over which host-pathogen coexistence is possible (Thrall, Biere, and Uyenoyama 1992; Thrall, Antonovics, and Hall 1992). In the models that follow, I include differential density-dependent regulation by assuming that it acts on the growth rate of the healthy class (e.g., via its effects on juvenile recruitment, or adult fecundity), but has no effect on the diseased class (death rates of diseased and healthy individuals are density independent).

Joint Numerical and Gene Frequency Dynamics

It is possible to develop theoretical models of host-pathogen systems that incorporate numerical and gene-frequency dynamics to differing degrees. Of particular interest are three special cases representing variation in gene frequencies only (no numerical dynamics), variation in numbers only (no gene frequency dynamics), and variation in both frequencies and numbers of diseased and healthy individuals/genotypes. For each, there may be both gene-frequency equilibrium and numerical equilibrium, there may be one or the other, or there may be neither. I also focus on particular equilibria, and ask whether they are approached in different ways given different assumptions about the genetic and numerical dynamics.

Unregulated vs. Regulated Populations

We begin by comparing the gene-frequency dynamics in a population that is unregulated (or where healthy and diseased individuals are proportionately regulated) with dynamics in a population where there is differential response of the two types of individuals to density. To again reflect the situation in the *Silene-Ustilago* system, we will consider the case where reproduction of (or recruitment into) the healthy class is a function of plant density, whereas the diseased class is unaffected by density. Throughout we assume that populations follow the "reciprocal yield law" characteristic of many plant populations (Harper 1977), and which reflects what is often termed pure "contest" competition in animal populations (Hassell 1975). This assumes that at very high densities the number of individuals emerging in the next generation asymptotes to some level, rather than decreases. We include population regulation by setting birth rate,

$$b = b_0/(b_1 N_t + 1),\qquad(7.5)$$

such that the carrying capacity,

$$K = (b_0 - d)/b_1 d,\qquad(7.6)$$

where b_0 = birth rate at low density

b_1 = a constant determining the intensity of density dependence

N_t = total population size (= $X_t + Y_t$)

d = death rate.

It can be shown that without some form of density dependence a stable genetic polymorphism is not possible. Typically the resistance gene either does not spread, or allele frequencies show cycles of ever-increasing amplitude. It can be seen from eqs. (7.3a–d) that the fitnesses of the resistant and susceptible genotypes do not depend on their relative frequencies, but only on the frequency of diseased individuals.

An example of an unregulated population is shown in figure 7.1a: here gene-frequency dynamics shows ever-increasing oscillations, leading eventually to extinction of one or the other genotype. However, when population regulation is included in the model but otherwise the parameters are unchanged, stable gene frequency (fig. 7.1b) and numerical dynamics are possible (fig. 7.1c).

Genetically Uniform vs. Genetically Variable Populations

At the simplest level, the importance of genetic variation in a host-pathogen system is obvious. Thus if a population is initially composed of only very susceptible individuals, we predict the certain and rapid demise of the whole population. However, the subsequent spread of a few resistant individuals can save such a population from extinction. As in all host-pathogen systems, it is obvious that in the presence of the disease, and with no fitness cost to resistance, the resistant genotype will spread in the population. However, if there is a cost to the resistance, then there is the

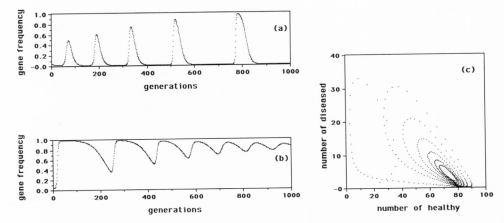

FIG. 7.1. (a) Gene frequency dynamics of a host-pathogen system assuming no popu-lation regulation. Graph shows change in frequency of the resistant allele over genera-tions. Model is based on eqs. (7.4a–c) (see text) with parameter values: $b1$, $b2 = 0.8$, 0.64; $d = 0.2$; $\beta1$, $\beta2 = 0.9$, 0.1. (b) Gene frequency dynamics when there is popula-tion regulation by density dependence acting on the healthy class. Model is based on eqs. (7.3) and (7.5) (see text) with parameter values as above, except that in addition $K1$, $K2 = 100$, 80. (c) Numerical dynamics for case (b), shown as a phase plane plot. Here and in all subsequent figures, starting numbers (or proportions) were 18 suscepti-ble, 1 resistant, and 1 diseased.

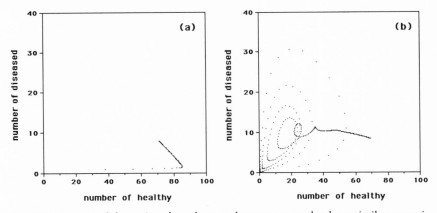

FIG. 7.2. Numerical dynamics of two host-pathogen systems that have similar numeri-cal equilibria, but where (a) the host population is genetically uniform, and (b) the host population is genetically variable. Model is based on eqs. (7.3) and (7.5) (see text) with parameter values: (a) $b = 4.4$; $d = 0.45$; $\beta = 0.5$; $K = 88.6$. (b) $b1$, $b2 = 0.9$, 0.6; $d = 0.4$; $\beta1$, $\beta2 = 0.9$, 0.35; $K1$, $K2 = 120$, 105.

possibility that the population may maintain a genetic polymorphism for resistance. Perhaps less obvious is the fact that even though the numerical equilibrium of a genetically variable host-pathogen system may be identical to one that is genetically invariant, the approach to equilibrium may be quite different. This is illustrated in figure 7.2, where it can be seen that in the case of a genetically uniform population, approach to equilibrium involves an initial overshoot on numbers of the healthy, but then a relatively direct and rapid convergence on the equilibrium (fig. 7.2a); in the genetically variable population, approach to numerical equilibrium is oscillatory and takes an extremely long time (fig. 7.2b). It is particularly noteworthy that in the genetically variable case, identical points in the phase plane can have quite different trajectories (also in fig. 7.1c). The reason is that the phase plane, as depicted, represents numerical composition and not genetic composition: thus populations represented by singular points on this plane can have different genetic compositions and can therefore change in different directions depending on this composition. To fully represent the dynamics, the phase plane should be in three dimensions, with gene frequency as a third axis: the dynamics shown in figure 7.2b would then be a path that takes the shape of an ever-narrowing helix. Any phase-plane analysis of a host-pathogen system will therefore be misleading if there is underlying genetic variation in the responses of the host and/or pathogen.

Demographic Form of the Resistance Cost

To illustrate how the ecological and demographic expression of the cost of resistance can affect the dynamics, we will compare the outcomes of three cases (fig. 7.3a–c). In all three cases the model parameters are identical except in the way the cost is included. Thus in figure 7.3a we assume that the cost affects the reproductive rate of the individuals, but has no effect on the carrying capacity. This might be a situation where the resistance gene affects reproductive output (e.g., more resistant individuals reproduce less), but that population size is a function of adult density and that density acts equally on resistant and susceptible individuals. In this case, a polymorphism cannot be maintained. The reason is that, as the resistance gene spreads, population size increases; at high densities, because resistant and susceptible genotypes have equal carrying capacities the fitness differential between them is reduced, and so resistant individuals further increase in frequency, again increasing population size, and so on. In figure 7.3b we assume that the reproductive rate of resistant and susceptible genotypes is the same, but resistant genotypes are more affected by increasing density. This might be a situation where the resis-

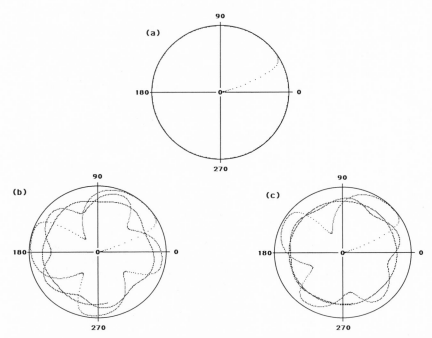

FIG. 7.3. Gene-frequency dynamics of a host-pathogen system assuming different forms of the resistance costs: (a) different reproductive rates, identical carrying capacities; (b) identical reproductive rates, different carrying capacities; (c) different reproductive rates and carrying capacities. Change in frequency of the resistant allele is plotted on polar coordinates, where distance from the center of the circle represents allele frequency (scale 0–1), and where generations are plotted in units of degrees starting at zero and increasing counterclockwise around the circle. Model is based on eqs. (7.3) and (7.5) (see text) with parameter values: for all cases, $d = 0.2$, $\beta 1$, $\beta 2 = 0.9$, 0.1. (a) $b1$, $b2 = 0.8$, 0.48; $K1$, $K2 = 100$. (b) $b1$, $b2 = 0.8$; $K1$, $K2 = 100$, 60. (c) $b1$, $b2 = 0.8$, 0.48; $K1$, $K2 = 100$, 60.

tance gene affects resource-use efficiency, such that resistant individuals require more resources for survival. Here a polymorphism can be maintained. However, under these conditions, imposing a fitness cost in terms of the reproductive rate is not without effect: the dynamics are again different when the cost affects not only the carrying capacity but also the reproductive rate of resistant genotypes (fig. 7.3c).

Another possibility is that the fitness cost is expressed in only one sex. This may well be the case in *Silene*, where resistant males have fewer flowers and flower earlier, but where no such trend is seen in females (Alexander 1989). If this is the case, then a polymorphism is maintained even though the resistant and susceptibles have identical carrying capaci-

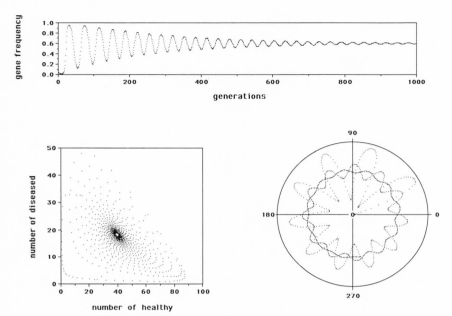

FIG. 7.4. Gene-frequency and numerical dynamics of a host-pathogen system assuming the resistance costs affect only male fecundity and the two resistance types have identical carrying capacities. Upper graph shows the change in frequency of the resistant allele over generations; the circle shows the same data plotted on polar coordinates (see legend to fig. 7.3 for explanation). Numerical dynamics is shown as a phase plane plot at the lower left. Model is based on eqs. (7.3) and (7.5), except that the resistance cost affects male fecundity. Parameter values: $b1$, $b2 = 0.8$; $d = 0.2$; $\beta1$, $\beta2 = 0.9, 0.1$; $K1$, $K2 = 100$; $m1$, $m2$ (male fecundities) = 1.0, 0.6.

ties and equal female fecundities (fig. 7.4). The reason is that when a highly fecund (and therefore susceptible) male is rare, it makes a relatively greater genetic contribution to the next generation than when it is common. This introduces an innate frequency dependence such that when susceptible individuals are reduced in frequency by the disease, their genetic contribution increases relative to that of resistant individuals. If there is in addition some cost in terms of overall carrying capacity, still different dynamics are evident, producing in this case stable limit cycles (fig. 7.5).

The interaction of venereal disease spread and male and female fitnesses is likely to be complex and have important repercussions for the evolution of mating systems and sexual selection. For example, low mating success (and greater resistance) of males may be genetically correlated with higher female fitness. Differential allocation to male and female

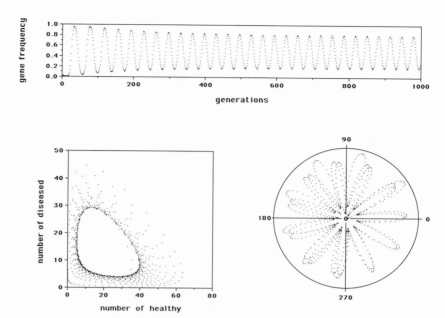

FIG. 7.5. Gene-frequency and numerical dynamics of a host-pathogen system assuming the resistance costs affect both male fecundity and carrying capacities of the two resistance types. For explanation and parameter values, see fig. 7.4, except here $K1$, $K2 = 100, 60$.

functions can itself maintain a genetic polymorphism: but the range of values over which polymorphisms in male and female mating success can be maintained may be much greater in the presence of venereal diseases.

Discussion

My goal in this chapter has been to emphasize the importance of including both genetic variation and numerical dynamics in models of host-pathogen interactions. Not only are expected equilibria and transient dynamics quite different if both components are included, but it becomes clear that purely genetic models based on frequencies of interacting components unwittingly subsume a frequency-dependent transmission mode (Jayakar 1970; Lewis 1981; Seger 1988). These models are therefore only directly applicable to venereal and vector-borne diseases. Several such models were originally developed with crop diseases in mind (Leonard 1977; Barrett 1988), and no doubt the investigators may be quite surprised to realize that they were in fact modeling venereal diseases! A number of authors, including some of the above, have pointed out that host-parasite interactions are likely to be both frequency and density dependent, but as far as I know only one paper has investigated this explicitly (May and Anderson 1983). In the other studies the focus has been on gene frequencies, but the models have been formulated with the additional assumption that disease spread is most likely when the susceptible genotype is most common, thereby generating frequency-dependent selection (Gillespie 1975; Clarke 1976).

The models presented here, while reflecting the essential features of a real host-pathogen system, are essentially heuristic; no attempt has been made to explicitly simulate the *Silene-Ustilago* system. Nor have the dynamical consequences of the models been exhaustively analyzed; the dynamical patterns are shown for illustrative purposes and are specific to the chosen parameter values. Nevertheless, there is no doubt that the overall conclusions regarding the importance of combining genetic and numerical dynamics are quite general. Indeed, the analyses presented here may understate the case because they may be uncharacteristic of other kinds of host-pathogen systems. Thus they assume no genetic variation in the pathogen, and therefore exclude the dynamical behaviors that can result from genetic specificity in host-pathogen interactions: adding pathogen variation can result in a highly complex gene-frequency dynamics (Seger 1988) and can lead to the maintenance of genetic variation in both host and pathogen in the absence of any externally imposed population regulation. The models that have been presented here also exclude the possibility of "within season" population dynamics, resulting because the generation time of the pathogen is much shorter than that of the host: such within-season epidemics can have severe effects on relative fitnesses and on total abundance (Gillespie 1975; May and Anderson 1983). In the *Silene-Ustilago* system, the anther-smut has a long latent period (from three to six weeks or even longer; Alexander, Antonovics and Kelly 1992) and therefore the number of pathogen generations per season is probably quite limited. May and Anderson (1983) modeled a pathogen with a fast life cycle relative to host (such that epidemics occurred on the host each season due to rapid density-dependent disease transmission). They also assumed there was host-pathogen specificity, such that each host-genotype could only be attacked by the corresponding pathogen genotype. This system exhibited very rich dynamics and could lead to frequency-dependent selection sufficiently severe to result in chaotic cycles of numbers and gene frequencies. If the actual abundances and not just the relative frequencies of the different genotypes were affected, gene frequency dynamics was even more likely to be chaotic. Although they did not explicitly investigate the consequences of disease-independent population regulation, they suggested that this may actually dampen any large or chaotic oscillations and result in more predictable dynamics. Many other host-pathogen systems are therefore likely to demonstrate a complex interplay between numerical dynamics and gene frequency.

Because of the multiplicity of factors that impinge on their dynamics, host-pathogen systems pose a serious challenge to the ecological geneticist. From a purely theoretical perspective, any model almost inevitably has to be an uncomfortable compromise between generality (simplicity and tractability) and realism (specificity). From an empirical perspective, understanding the forces that impinge on the dynamics of host-pathogen systems requires detailed information of the kind that is not easily gleaned by, say, a simple population survey. Were nature so simple and convenient as to be describable by these basic models, estimation of the model parameters would still present formidable problems. For example, in the *Silene-Ustilago* system considered here, parameterization of the models would require knowledge of genetic variation in the recruitment rates, the death rates, and the degree and type of density-dependent population regulation. We would also need to know about genetic variation in the disease transmission coefficient. Even calculating a disease transmission coefficient in the absence of genetic variation is fraught with difficulties: it may appear easy to calculate this from the rate at which healthy individuals become diseased and the frequency of the disease in the previous time interval (year), but there are further hidden assumptions. Most importantly we have to assume that the disease transmission coefficient is constant and does not vary with population composition, or with the frequency and density of diseased individuals. None of these assumptions can be validated by a descriptive study of one population at one time interval. If the population is itself genetically variable for resistance, then any predicted trajectory based on purely ecological data may be very misleading precisely because the genetic composition of the population will change.

There are two general lessons to be learned from consideration of the interplay of numerical and genetic processes in host-pathogen systems. The first is that any pathogen control strategy based on modeling population behavior needs to be based on both genetic and ecological information. The second is that if we are to understand disease in natural populations, then two divergent but complementary strategies are available to us. One strategy is to study a particular system, but this perforce carries the expectation that we will need a wealth of genetic and demographic detail to gain a thorough understanding. The other strategy is to develop generalized expectations about the role of pathogens in maintaining species diversity in communities and genetic diversity within species, and then use comparative and experimental approaches to test these expectations. I have suggested that there is a real place for such a holistic approach to understanding biotic interactions in natural populations, and that this should perhaps be encompassed by a new discipline termed "community genetics" (Antonovics 1992). As we have seen from the examples in this chapter, biotic interactions in communities are likely to be, if not dominated, then at least greatly influenced by genetic factors. Conversely, there is now increasing evidence that pathogens are likely to be important forces in the evolution of mating systems (Levin 1975; Hamilton 1980; Schmitt and Antonovics 1986; Lively 1987; Lively et al. 1990). Because of this close interplay of ecological and genetical processes, it is very likely that features such as the mating system or the recombination frequency, which in the past have been viewed as only affecting long-term evolutionary potential and were of little direct interest to the ecologist, may themselves influence species abundances and distributions, via their effect on disease and pest incidence (Burdon and Marshall 1981). The trend in ecology toward a consideration of biotic rather than abiotic forces in determining species abundances and distributions will increasingly engage the expertise and interest of the ecological geneticist.

Acknowledgments

I wish to thank Peter Thrall for overseeing my algebra, and Helen Alexander for her insights and collaboration which stimulated this study. Support was provided by NSF grant BSR-8717664, a Guggenheim Fellowship, and a Royal Society Visiting Research Fellowship at the Biology Department, Imperial College at Silwood Park, University of London.

References

Alexander, H. M. 1989. An experimental field study of anther-smut disease of *Silene alba* caused by *Ustilago violacea*: Genotypic variation and disease incidence. *Evolution* 43: 835–847.

Alexander, H. M. 1990. Epidemiology of anther-smut infection of *Silene alba* caused by *Ustilago violacea*: Patterns of spore deposition and disease incidence. *J. Ecol.* 78: 166–179.

Alexander, H. M., and J. Antonovics. 1988. Disease spread and population dynamics of anther-smut infection of *Silene alba* caused by the fungus *Ustilago violacea*. *J. Ecol.* 76: 91–104.

Alexander, H. M., J. Antonovics, and A. W. Kelly. 1993. Genotypic variation in plant disease resistance: Physiological resistance in relation to field disease transmission. *J. Ecol.* 81: 325–333.

Anderson, R. M., and R. M. May. 1978. Regulation and stability of host-parasite population interactions. I. Regulatory processes. *J. Animal Ecol.* 47: 219–247.

Antonovics, J. 1992. Towards community genetics. In *Plant Resistance to Herbivores and Pathogens: Ecology, Evolution and Genetics*, ed. R. S. Fritz and E. L. Simms, pp. 426–449. University of Chicago Press, Chicago.

Antonovics, J., and H. M. Alexander. 1989. The concept of fitness in plant-fungal pathogen systems. In *Plant Disease Epidemiology*, ed. K. J. Leonard and W. E. Fry, vol. 2, pp. 185–214. McGraw-Hill, New York.

Antonovics, J., and H. M. Alexander. 1992. Epidemiology of anther-smut infection of *Silene alba* caused by *Ustilago violacea*: Patterns of spore deposition in experimental populations. *Proc. Roy. Soc. London B* 250: 157–163.

Baker, H. G. 1947. Infection of species of *Melandrium* by *Ustilago violacea* (Pers.) Fuckel and the transmission of the resultant disease. *Ann. Botany* 11: 333–348.

Barrett, J. A. 1988. Frequency dependent selection in plant-fungal interactions. *Phil. Trans. Roy. Soc. London B* 319: 473–483.

Burdon, J. J. 1987. *Disease and Plant Population Biology*. Cambridge University Press, Cambridge, U.K.

Burdon, J. J., and G. A. Chilvers. 1977. The effect of barley mildew on barley and wheat competition in mixtures. *Australian J. Botany* 25: 59–65.

Burdon, J. J., and D. R. Marshall. 1981. Biological control and the reproductive mode of weeds. *J. Appl. Ecol.* 18: 649–658.

Clarke, B. 1976. The ecological genetics of host-parasite relationships. In *Genetic Aspects of Host-Parasite Relationships*, ed. A.E.R. Taylor and R. Muller, pp. 87–103. Blackwell, Oxford.

Clay, K. 1991. Parasitic castration of plants by fungi. *Trends in Ecol. and Evol.* 6: 162–166.

Day, P. R. 1974. *Genetics of Host-Parasite Interaction*. Freeman, San Francisco.

De Wit, C. T. 1960. On competition. *Verslagen van landbouwkundige onderzoegen*, no. 66–8. Wageningen.

Fritz, R. S., and E. L. Simms. 1992. *Plant Resistance to Herbivores and Pathogens: Ecology, Evolution and Genetics*. University of Chicago Press, Chicago.

Getz, W. M., and J. Pickering. 1983. Epidemic models: Thresholds and population regulation. *Amer. Nat.* 121: 892–898.

Gillespie, J. H. 1975. Natural selection for resistance to epidemics. *Ecology* 56: 493–495.

Hamilton, W. D. 1980. Sex versus non-sex versus parasite. *Oikos* 35: 282–290.

Harper, J. L. 1977. *Population Biology of Plants*. Academic Press, London.

Hassell, M. P. 1975. Density dependence in single species populations. *J. Animal Ecol.* 44: 283–295.

Hassel, M. P. 1978. *The Dynamics of Arthropod Predator-Prey Systems*. Princeton University Press, Princeton, N.J.

Jayakar, S. D. 1970. A mathematical model for interaction of gene frequencies in a parasite and its host. *Theor. Pop. Biol.* 1: 140–164.

Leonard, K. J. 1977. Selection pressures and plant pathogens. In *The Genetic Basis of Epidemics in Agriculture*, ed. P. R. Day, pp. 207–222. Annals of the New York Academy of Sciences 287.

Levin, D. A. 1975. Pest pressure and recombination systems in plants. *Amer. Nat.* 109: 437–451.

Lewis, J. W. 1981. On the coevolution of pathogen and host. I. General theory of discrete time coevolution. *J. Theoret. Biol.* 93: 927–951.

Lively, C. M. 1987. Evidence from a New Zealand snail for the maintenance of sex by parasitism. *Nature* 328: 519–521.

Lively, C. M., C. Craddock, and R. C. Vrijenhoek. 1990. Red Queen hypothesis supported by parasitism in sexual and clonal fish. *Nature* 344: 864–866.

May, R. M., and R. M. Anderson. 1983. Epidemiology and genetics in the coevolution of parasites and hosts. *Proc. Roy. Soc. London B* 219: 281–313.

Schmitt, J., and J. Antonovics. 1986. Experimental studies of the evolutionary significance of sexual reproduction. IV. Effect of neighbor relatedness and aphid infestation on seedling performance. *Evolution* 40: 830–836.

Seger, J. 1988. Dynamics of some simple host-parasite models with more than two genotypes in each species. *Phil. Trans. Roy. Soc. London B* 319: 541–555.

Thrall, P. H., A. Biere, and M. Uyenoyama. 1992. The population dynamics of host-pathogen systems with frequency dependent disease transmission. *Amer. Nat.* (to be submitted).

Thrall, P. H., J. Antonovics, and D. W. Hall. 1992. Host and pathogen coexistence in vector-borne and venereal diseases characterised by frequency dependent disease transmission. *Amer. Nat.* (in press).

Vanderplank, J. E. 1984. *Disease Resistance in Plants.* Academic Press, London.

Wolfe, M. S., and C. E. Caten. 1987. *Populations of Plant Pathogens: Their Dynamics and Genetics.* Blackwell, Oxford.

JANIS ANTONOVICS

8

Ecological Genetics of Metapopulations:
The *Silene-Ustilago* Plant-Pathogen System

(with Peter Thrall, Andrew Jarosz, and Don Stratton)

Introduction

There has been a long-standing recognition that the numerical and gene-frequency dynamics of natural populations will be affected both by processes that occur locally within populations as well as by factors that affect the interconnectedness and persistence of populations on a more regional scale (Hutchinson 1953; Wright 1943). Historically, and perhaps also for heuristic and practical reasons, primacy has been given to a study of processes within populations. Recently, however, a number of issues have forced greater attention on population processes at a regional scale. These issues include practical problems such as predicting the consequences of habitat fragmentation (Burkey 1989; Wilcox 1990), as well as conceptual advances such as the demonstration that populations which cannot persist locally may still show equilibrium persistence on a regional scale (Levins 1969; Caswell 1978; Hanski and Gilpin 1991) and that genetic structure can be greatly affected not just by migration among extant populations but by colonization and extinction processes (McCauley 1991). Above all there has been an increasing recognition that in nature no population is an isolated entity, and that this reality needs to be quantified and its consequences need to be explored. Metapopulations (systems of interconnected populations) are more likely to be the rule, not the exception.

The major theoretical and empirical issue at the heart of discussions about metapopulations is how the global behavior of a system of interconnected populations can be understood in terms of the local dynamics of individual populations, their degree of synchrony, and the connectedness among them (Hanski and Gilpin 1991). This issue can be broken down into two interacting questions. The first question is the degree to which interconnectedness of populations affects local dynamics. The second question is the degree to which local dynamics in turn affects outcomes on a more regional scale. These questions are problematical because although "a population" may be easily defined in theory or by the confines of an experimental container, in a mosaic patchwork such as is found in nature there is no easily agreed-upon definition of a population. The ecological, genetic, and evolutionary criteria that have been used for delimiting a population may give quite different dimensions (Antonovics and Levin 1980; Uyenoyama and Feldman 1980). One extreme population structure would be exemplified by situations where "populations" are represented by cells within a grid superimposed upon a continuum of individuals and where there is a high degree of interpatch migration, while the other extreme would be exemplified by "populations" that are on quite isolated islands or in distinct habitat patches and where migration rate per generation is rare. The term "metapopulation" has often been restricted to the latter situation, and we use it in that sense here but with the clear expectation that actual population structures may be somewhere between these extremes, or represent some combination of them (e.g., the "core-satellite" hypothesis; Hanski 1982). Much of the challenge of the empirical study of metapopulations is simply defining the actual structure and connectedness of the component study units (Harrison 1991; Taylor, A. D. 1991).

In nature, the interaction between local population dynamics and population connectedness confounds their easy dissection. From a simple description of the numerical dynamics of a local population or patch, one does not know (unless the migrants can be explicitly identified) whether it is being influenced by connectedness. For example, local instability may be undetectable because population extinctions do not occur because of immigration from other nearby populations (the "rescue effect"; Brown and Kodric-Brown 1977). Experimental approaches are necessary which manipulate the frequency of migration/colonization events among component populations. Studies in the laboratory have shown that persistence of a set of populations can be increased by some connectedness among them (Huffaker 1958; Pimentel et al. 1963; Takafuji 1977), and although experimental studies under natural or seminatural conditions are much rarer, they have generally given the same result (Kareiva 1984, 1987; Sabelis and van der Meer 1986; Sabelis and Laane 1986).

The subject of metapopulations has also been approached from a theoretical point of view by subsuming the details of local dynamics in order to achieve generalizations about the more global behavior of the metapopulation as a whole. The pioneering approach of theoretical island biogeography, for example, assumed a source population (or "refuge"), and asked about the effect of migration on population extinction and colonization, while ignoring the details of within-island dynamics (MacArthur and Wilson 1967). The classical metapopulation model of Levins (1969) also assumed no within-population dynamics other than "presence" after colonization and "absence" after extinction. Models which incorporate local population dynamics ("structured" metapopulation models; Hanski 1991) become much more difficult to analyze theoretically but can lead to qualitatively different results, such as the requirement that there be a minimum fraction of occupied populations for the metapopulation to increase (Gyllenberg and Hanski 1992) or results such as nonmonotonic changes in metapopulation size with increasing migration (May and Anderson 1990). It is therefore important to know under what conditions local dynamics matters; in some instances equilibrium may be approached rapidly and local dynamics can be subsumed with little loss of accuracy (Verboom and Lankester 1991).

Even where structured metapopulation models have been developed, they have usually assumed that migration is important in the colonization process, but that subsequent migrants have little effect on local dynamics (Diekmann et al. 1988; Hastings and Wolin 1989; Sabelis et al. 1991). However, the colonization process involves not only demographic stochasticities, but also genetic stochasticities. Thus, rare migrants subsequent to a colonization event may have little immediate effect on local numerical dynamics, but may have a large impact on the genetic dynamics, which in turn may affect population abundances. For example, migrants may reduce the level of inbreeding that might otherwise ensue if there were only one or a few colonists; or they may introduce genotypes (e.g., resistance or virulence types) that may have been absent in the founding population.

In this discussion we use the *Silene-Ustilago* system to illustrate several important aspects of the ecology and genetics of metapopulation systems. First, we present the results of a long-term study of populations of *Silene alba* and *Ustilago violacea* spanning a broad regional scale that includes several hundred populations; we examine the rates of population turnover, and the degree of interconnectedness among the component populations. Second, we present some initial results from empirical studies and computer simulations which address the question of how genetic structure of host-pathogen systems might affect metapopulation dynamics.

The Silene-Ustilago Host-Pathogen System

Silene alba or white campion (= *Silene latifolia*, *Melandrium album*; hereafter termed *Silene*) is a short-lived perennial herb, commonly found in ruderal habitats throughout the northern regions of the United States and in upland areas farther south (McNeill 1977). It is easily grown and flowers in about six weeks under long days in a growth chamber; it is dioecious and easily crossed. *Ustilago violacea* or anther-smut fungus

(= *Microbotryum violacea*; hereafter termed *Ustilago*) causes both male and female plants to produce anthers that carry fungal spores instead of pollen; diseased females produce a sterile, rudimentary ovary. The disease is systemic, and after the initial stages of infection, all flowers become diseased and the plant is sterilized, although plants becoming diseased toward the end of the growing season may recover. Diseased flowers are easily recognized by their dark, smutted centers. The spores are transmitted by pollinators and to a limited extent by passive scattering (Alexander 1990a). Spores germinate on the host and undergo meiosis to produce haploid sporidia which multiply by budding. Fusion of sporidia of opposite mating type produces a heterokaryotic infection hypha.

Silene shows genetic variation in resistance to the disease in field experiments (Alexander 1989; Thrall 1993). A single population may contain mixtures of highly susceptible and highly resistant individuals, but the precise inheritance of resistance is unknown. Resistance in the field is correlated with resistance in the greenhouse, except that differences in field susceptibility are additionally related to differences among host clones in flowering time (Alexander, Antonovics, and Kelly 1993).

The fungus appears to be genetically much less variable. Artificial inoculations revealed no differences in virulence among a limited number of lines isolated from one population (Alexander, Antonovics, and Kelly 1993). A search for other genetic markers revealed only very limited variation for allozymes and RFLP's (Stratton, unpublished). More recently the use of randomly amplified polymorphic DNA (or RAPDs: Williams et al. 1990; Martin et al. 1991) has been used to successfully differentiate twelve sporidial lines from one population; crosses among these isolates have confirmed the Mendelian inheritance of all markers tested to date (Oudemans, unpublished). However, we do not know to what extent such genetic variation at the molecular level is reflected by variation in virulence.

A major stimulus for our research has been the expectation that *Ustilago*, because it is pollinator transmitted, is likely to show frequency-dependent disease transmission, a transmission mode that is characteristic of venereal or vector-borne diseases (Getz and Pickering 1983). Frequency-dependent transmission occurs when the probability of a healthy individual becoming diseased is a function of the frequency or fraction of individuals in the population that are diseased, rather than a function of overall plant density (see chap. 7). Such transmission might be expected whenever disease vectors (mates or pollinators) move further to compensate for increased spacing among individuals at lower population densities. Antonovics and Alexander (1992) showed that spore deposition increased with frequency but not with density of diseased individuals in experimental populations of diseased and healthy *Silene* exposed to natural pollinators. Simple models show that while normal density-dependent transmission processes require a threshold density for disease spread, and can readily regulate population size, frequency-dependent transmission modes can lead to either extinction of both host and pathogen or purging of the disease from the population (Getz and Pickering 1983; Antonovics 1992). Host-pathogen coexistence is possible if there is density-dependent host population regulation, for example, by resource limitation, and such coexistence is most likely if density acts more severely on the healthy class than on the diseased class (Thrall, Antonovics, and Hall 1993). Greater density-dependent limitation on the healthy class may be expected in cases (as in pollinator-transmitted and other venereal disease systems) where the disease is transmitted in the adult phase, and density-dependent population regulation occurs primarily in the juvenile (or seedling) stage.

Models that incorporate genetic variation in host resistance show that if there are costs to disease resistance, plant populations can be polymorphic for genotypes that in a monomorphic state would either lead to host-pathogen extinction or failure of the disease to establish (Antonovics 1992; see chap. 7). This leads to the prediction that extremes of resistance and susceptibility can be maintained in one population, a prediction that is supported by the results of Alexander (1989).

Metapopulation Dynamics: Natural Populations

To study a system of interconnected populations, it is necessary operationally to define a component "population" and the universe that constitutes the "metapopulation." We have chosen as our "metapopulation" a 25 × 25 km area in the vicinity of Mountain Lake Biological Station in southwestern Virginia (fig. 8.1). Our "populations" are defined as those individuals (healthy and diseased) in a 44-yard (c. 40 m) segment of roadside. Over the past three years, we have studied nearly 150 km of roadside, encompassing about seven thousand roadside segments. In any one season, over 300 of these have *Silene* growing in them, and about seventy have at least some diseased individuals. We have found the host and the disease in the immediate area outside our census region as well as up to 50 miles away. We therefore have no reason to believe that our study area is in any way exceptional.

Most metapopulation studies have focused on populations found in discrete habitat types, so that the potential sites available to them can be delimited geographically (Rey and Strong 1983; Harrison 1991; Peltonen and Hanski 1991). However, in our system potential sites are not easily defined, and we therefore describe metapopulation processes in terms of the numbers of individuals in each of the roadside segments. In the region

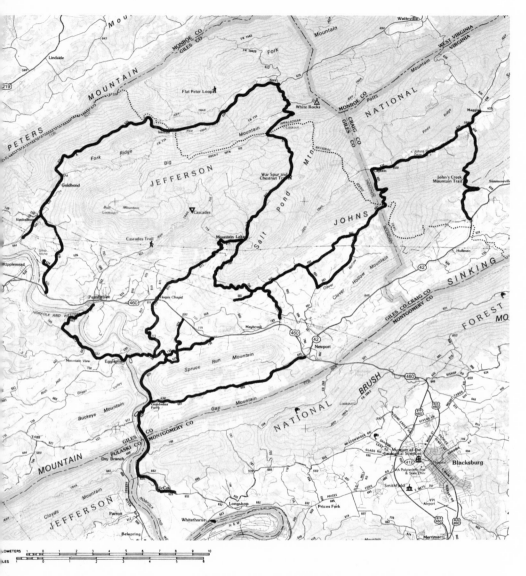

FIG. 8.1. Map showing the roads (thick black lines) included in the annual census of *Silene alba* populations. (From Virginia Atlas & Gazetteer, copyright DeLorme Mapping Co. Reproduced with permission)

of Virginia (Giles County) where we have been studying the *Silene-Ustilago* system, land use is mostly grazed pastures and woodlands and *Silene* is largely confined to roadsides. The scale of 44 yards allows us easily to locate a segment using a car odometer and to study a large number of roadside segments covering a wide area while having, at least potentially, a substantial number of individuals within each segment. Each segment also encompasses at least one, but not numerous, fungal and host genetic neighborhoods. Studies of local spore dispersal showed that at 10 m from a point source, only a third of the flowers had spores deposited on them relative to flowers close to the source; these flowers had only about a tenth of the number of spores as flowers at the source (Alexander 1990a). Because the spores are pollinator transmitted, pollen dispersal is likely to be of the same order of magnitude. On the other hand, seed dispersal is much more limited, with most seeds falling within 2 m of the seed source.

We census the populations twice a year. The main census is carried out in early June, with a secondary census at the end of July to recheck segments that showed critical state transitions (i.e., where plants or disease were not seen in the current year but were present the previous year). During the main census, each roadside segment is marked using local landmarks (unusual trees, telephone posts, buildings, etc.) so it can be relocated precisely from one year to the next. We record each side of the road separately because the sides often differ in aspect, vegetation, disturbance history, and disease incidence. We also take pains to avoid becoming disease dispersal agents ourselves: the census is carried out prior to midmorning closure of the flowers so diseased plants can be easily identified without touching the flowers. Data for two years from a representative one-mile section of road are shown in figure 8.2.

The results of the first three years of the census have shown that there is a high turnover rate of populations (table 8.1a), so much so that it is possible to calculate "vital statistics" for whole populations (table 8.1b) in a manner that is similar to what is normally done for individuals within populations. These results show that over the three-year period 1988–1990, the colonization rate of the disease was greater than the extinction rate. These results suggest that the disease is spreading in this region of Virginia, a trend that is confirmed by comparison with a smaller number of populations in the census area that were studied in some detail in 1984 (Alexander 1990b). For the host, the colonization rate exceeded the extinction rate in 1988–1989, but the reverse was true in 1989–1990. Population extinction rates were "size dependent," being higher in small than in large populations (table 8.2a). Although there has been much speculation about "minimum viable population" sizes in the conservation literature, empirical data which bears on this issue is almost nonexistent (Wilcox 1990). Extinction rates of small populations were less when they

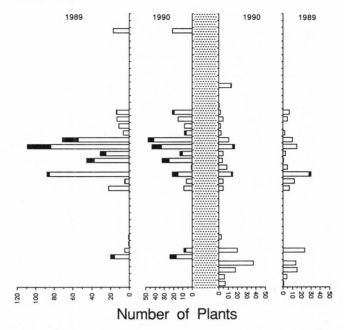

Number of Plants

FIG. 8.2. Illustration of data from a representative section of roadway, showing the results of two successive censuses, 1989 and 1990. The road is symbolized by the shaded region, with data for the left- and right-hand side of the road shown on either side. Each tick on the vertical axes delimits a 44-yd roadside segment (total vertical scale = 1 mile); the horizontal bars show the numbers of healthy (open) and of diseased (dark) individuals in each roadside segment in each year.

were in proximity to other populations (table 8.2b), suggesting the possibility that they might have been "rescued" from extinction by migrants from nearby populations (although clearly one cannot exclude the possibility that populations near each other are in favorable habitats).

Interesting patterns have emerged also with regard to disease incidence. Populations are more likely to be diseased when large (fig. 8.3). This again suggests a number of possible scenarios: large populations may attract more long-distance pollinators/spore vectors; or if there is a demographic cost to resistance, large populations may be more disease susceptible; or larger populations may be simply older populations that have had a greater chance of becoming diseased. Patterns similar to these have been seen with *U. violacea* on *Viscaria vulgaris* in Europe (Jennersten et al. 1983). Given that a population is diseased, small populations generally have a higher percentage of disease than large populations (fig. 8.4). This is predicted from frequency-dependent but not density-dependent disease-transmission processes; a single diseased individual in

TABLE 8.1
State transitions and metapopulation vital statistics for the occurrence of *Silene* and *Ustilago* in roadside segments, 44 yd long, in Virginia.

a. State transitions for all segments, where $t - 1$ indicates status the previous year, and t the status in the subsequent year. The category "diseased" includes partially diseased populations. The top number in each cell represents the transition from 1988 to 1989, and the bottom number the transition from 1989 to 1990. (More segments were included in the later censuses.)

	Year t		
Year t − 1	No Plants	Healthy	Diseased
No plants	5,254	139	12
	6,506	70	10
Healthy	19	161	23
	86	240	22
Diseased	3	5	40
	3	18	53

b. Metapopulation vital statistics showing rates of colonization and extinction of healthy populations, and rate of colonization and extinction of the disease (diseased populations). Rates are calculated as the number of new/extinct populations per existing healthy or diseased population per year. Population disease transmission rate is the probability of a healthy population becoming diseased per existing diseased population per year.

	1988 to 1989	1989 to 1990
A. Healthy populations:		
Colonization rate	0.60	0.19
Extinction rate	0.09	0.25
B. Diseased populations:		
Colonization rate	0.69	0.42
Extinction rate	0.16	0.30
C. Interaction:		
Rate at which healthy populations become diseased	0.10	0.07
Population disease transmission rate	0.002	0.001

Note: Here, and in subsequent tables, segments on opposite sides of the road are considered separately; occupied segments are termed populations.

TABLE 8.2

Extinction rate of healthy *Silene* populations as a function of their size and distance from other populations. Data are based on a comparison of the 1989 and 1990 censuses.

a. Relationship between population size and extinction rate (fraction of populations of a given size that went extinct between 1989 and 1990).

Population Size	No. of Populations	Extinction Rate
1	78	0.513
2–3	68	0.221
4–7	68	0.279
8–15	62	0.113
16–31	31	0.097
32–63	18	0
64–127	12	0
128–255	8	0
>255	3	0

b. Relationship between extinction rate of small populations (1–5 individuals) and their distance from the nearest population. Distance is measured in terms of number of roadside segments (each segment = 44 yd), and the nearest neighbor criterion considers populations on both sides of the road; populations directly opposite are considered to be one roadside segment away.

Distance (Roadside Segments) from Nearest Population	No. of Populations Extinct	Extinction Rate
1	80	0.600
2–3	24	0.667
4–7	24	0.682
8–15	26	0.654
>15	28	0.790

Note: Regression of distance versus weighted arcsine-transformed extinction rate: $Y = 0.845 + 0.0415X$, $p < 0.043$.

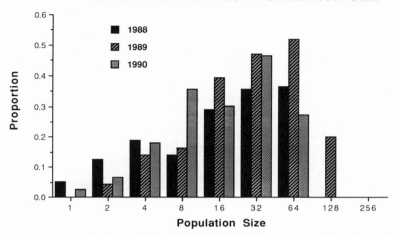

PROPORTION OF POPULATIONS DISEASED AS FUNCTION OF HOST POPULATION SIZE

FIG. 8.3. Proportion of *S. alba* populations that are diseased in relation to total population size in 1988, 1989, and 1990. The value for population size on the horizontal axis represents the lower value of the range for each size class.

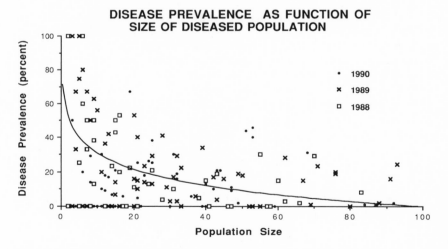

DISEASE PREVALENCE AS FUNCTION OF SIZE OF DISEASED POPULATION

FIG. 8.4. Disease prevalence as a function of population size. To discount the bias that would be introduced by the fact that all diseased populations have to have at least one diseased individual, the disease prevalence (percentage of diseased individuals in the population) is calculated using the transformation $100 * (D-1)/(T-1)$, where D and T are number diseased and total number, respectively. Formula for the fitted curve is: $Y = 61.5 - 30.9 * \log(X)$; $R^2 = 0.26$.

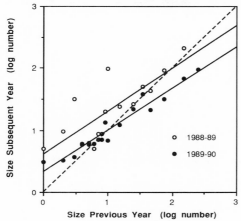

POPULATION SIZE IN THE SUBSEQUENT YEAR AS FUNCTION OF SIZE THE PREVIOUS YEAR

FIG. 8.5. Relationship between size of healthy populations and their size in the subsequent year. Each point represents the average of a population size class grouped such that numbers in each class are not less than five. The dotted line represents no change in population size over successive years; the solid lines are regressions for 1988–1989 (slope = 0.70, R^2 = 0.64) and 1989–1990 (slope = 0.67, R^2 = 0.95).

a small population will have a higher initial frequency than one in a large population, and the disease will therefore spread faster. With density-dependent disease transmission, smaller (and probably less dense) populations would have lower rates of disease increase.

Comparison of population size changes over successive years shows that the proportionate rate of increase of populations is less if the population size in the previous year was larger (fig. 8.5; table 8.3), or if there was a higher proportion of diseased individuals in the population (table 8.3).

TABLE 8.3
Effect of disease frequency and total population size on population growth rate of diseased populations between 1988 and 1989, and between 1989 and 1990. The table shows regression coefficients from a multiple regression of the natural log of change in size between years as a function of total size and disease frequency of the population in the previous year.

	Disease Frequency at Time t	Population Size at Time t	Model R^2
1988 to 1989	−0.35 n.s.	−0.42**	0.21
1989 to 1990	−0.52*	−0.36**	0.14

Significance levels: * = < 0.05, ** = < 0.01.

FREQUENCY vs DISTANCE FOR NEW HEALTHY POPULATIONS

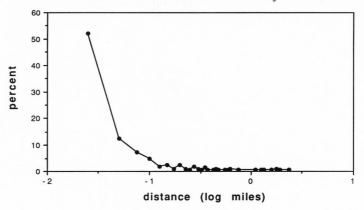

FREQUENCY vs. DISTANCE FOR NEW DISEASED POPULATIONS

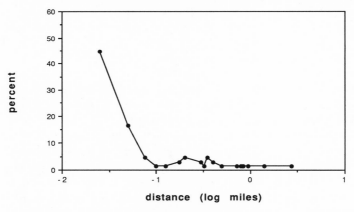

FIG. 8.6. Percentage of newly established populations that establish at a given distance from the population nearest to it in the previous year. *Top*: Host populations. *Bottom*: Pathogen populations. Combined data for 1989 and 1990. *Note*: distance measures are in log miles such that –1 = 176 yd, and 0 = 1 mile.

These results suggest density-dependent population regulation and a negative impact of the pathogen on the host population growth rate. The establishment of both the host and the disease depended on the distance from the nearest source population (fig. 8.6). While most new populations were founded close to existing ones, new populations often arose at considerable distances from preexisting ones. These data argue strongly for population interconnectedness even over substantial distances. Although we can never be absolutely sure that *S. alba* is not present in some intermediate nonroadside population (it is impossible to look in every-one's backyard!), such long-distance establishment strongly suggests oc-casional long-distance dispersal of seeds and, in the case of the pathogen, spores. The mechanism of long-distance transport is unknown; because roadsides are frequently mowed and disturbed, we suspect that humans and their machines may be important long-distance vectors. In the host, it is likely that the seeds can stay viable in the soil, and "long-distance" colonization may also reflect the earlier presence of populations which still persists in the seed bank; studies are under way to examine this possi-bility. Long-distance spore dispersal may also be affected by particularly vagile pollinators such as hawk-moths which have been observed to visit *Silene*.

Metapopulation Dynamics: Simulation Studies

In this section we present the results of computer simulations designed to examine whether the metapopulation processes of extinction and coloni-zation can maintain host-pathogen coexistence even in situations where in any one local population the host is driven to extinction by the patho-gen (and also resulting in the extinction of the pathogen itself). The simu-lations follow the general protocols used by Caswell (1978). We consider an array of segments, where each segment is characterized by the presence or absence of healthy or diseased individuals. The probability of a host or pathogen arriving in any segment is independent of distance, and there-fore is simply proportional to the number of hosts or pathogens summed over all segments, multiplied by a probability of per propagule establish-ment within a segment. Establishment within an already occupied site is ignored, and pathogens can establish in segments only where the host is already present. There is also an externally imposed likelihood of distur-bance resulting in disease-independent extinction. We assume that inter-nal dynamics are simple (disease incidence is a linear function of time since disease arrival, and host abundance is inversely proportional to disease incidence); postcolonization migrants are assumed to have no ef-fect on local dynamics. We also assume there may be a minimum time ("finding time") before the disease can establish itself in a newly founded healthy population.

The results of this model (fig. 8.7) show that, given estimates of extinc-tion and establishment rates based on our first two seasons of census data, host and pathogen coexistence can occur even when within-popula-tion dynamics leads to local extinction. As with other similar studies, a critical component was finding time (Murdoch 1977; May 1978; Nisbet and Gurney 1982). With no delay in pathogen arrival, host and pathogen

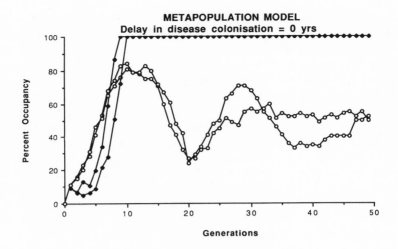

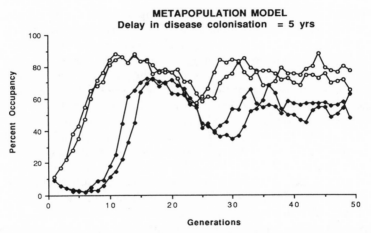

FIG. 8.7. Results from a computer simulation of metapopulation dynamics of one hundred interconnected segments. The two simulations differ in the time taken for the disease to colonize a previously established healthy population: (*top*) simultaneous colonization is possible; (*bottom*) colonization is not possible before five years. Two replicate runs are shown for each simulation. Vertical axis shows the percentage of segments occupied by populations regardless of whether they are diseased or healthy (open circles) and the percentage of these occupied segments that are diseased (solid circles). The models assume that the extinction time due to disease = 10 years; disease independent extinction rate = 0.08 per year; plant population productivity = 0.6 potential colonizations per year; pathogen population productivity = 10 potential colonizations per year.

continued to coexist within the metapopulation, but all populations became diseased (fig. 8.7, top). If there was a five-year delay in disease arrival, then only about half the populations became diseased (fig. 8.7, bottom).

The Genetic Component in Among-Population Processes

Although metapopulation models incorporating genetic processes have shown large effects of extinctions and colonizations on overall heterozygosity and genetic substructuring (Gilpin 1991; McCauley 1991; and Slatkin, chap. 1, this volume), very few have examined how genetic processes may influence numerical dynamics (for an exception, see Frank 1991). Clearly, the fate of a newly colonized host or pathogen population will depend in part on the genotypes of the initial colonists, and on how subsequent migrations augment the genetic variation present among these colonists.

Genetic variability is likely to affect metapopulation dynamics whenever new populations are founded by relatively few individuals that represent only a subsample of the global genetic diversity. This will have two major consequences: it will greatly affect the dynamics within each of the component populations, and it will generate the conditions for interdemic group selection. Within populations, the presence of only a few founders is likely to result in inbreeding because of mating among relatives. The reduction in fecundity and survival that results from inbreeding depression may contribute to the extinction of newly colonized populations through chance events, or it may have an impact on the subsequent growth rate of the population. In addition, if only a subsample of the global genetic diversity is represented in the founding population, then the evolutionary and numerical responses of the component populations may come to vary dramatically. This will be particularly important in host-pathogen systems where the genetics of the host and the pathogen can play an important role in numerical abundance (see chapter 7).

At the interpopulation level, if population turnover is high, whether by stochastic effects or because of a pathogen or predator, colonizers are likely to be from populations that show the greatest persistence and numerical abundance. This sets the stage for evolution of traits that increase group productivity or reduce variation in group size (Gilpin 1975; Wilson 1983). Therefore, assessing numerical changes caused by the interaction of ecological and evolutionary processes is critical to an evaluation of the likelihood of group selection in real-world systems.

We have many reasons to believe that the dynamics of the *Silene-Ustilago* metapopulation system may be affected by genetic processes, even though at present we have direct evidence only of genetic variation in resistance in the host. First, it is likely that the pool of potential colonists will be affected by the history of disease in the source populations. Thus colonists from populations with a history of disease may well be resistant, while those from populations without a history of disease will be susceptible. The potential impact of this is illustrated by our results from some experimental populations that we have set up to study the long-term dynamics of the *Silene-Ustilago* system. In 1990 we established sixteen "population cages" in the vicinity of Mountain Lake Biological Station. These cages consist of fenced (to prevent deer grazing) areas in which there are pots with healthy individuals, pots with diseased individuals, and pots that have soil but no plants and into which there can be seedling recruitment. The cages were established using a range of initial frequencies of diseased and healthy plants, but more critically, they were established from the progeny of only three pairs of parents. These parents were either susceptible or resistant individuals as determined in the experiments of Alexander (1989). The first year's results from these experiments dramatically illustrate how disease spread can be influenced by initial population composition. In those populations started from the progeny of susceptible individuals, 15.2% of the flowering plants ($n = 224$) became diseased in the first summer; however, in those populations started from the progeny of resistant individuals, only 0.6% of the flowering plants ($n = 181$) became diseased.

We can further illustrate the potential impact of genetics on the numerical dynamics of local populations using computer simulations (table 8.4; see also chap. 7, this volume). For illustration, we again restrict ourselves to cases where there is genetic variation in the host but not in the pathogen. The results are rather obvious but nonetheless dramatic. First, populations with different underlying genetic structures but similar numerical equilibria may approach these equilibria in quite different ways (fig. 8.8a,b; see also chap. 7, this volume). Given the stochastic nature of colonization events, colonizers are likely to differ in their genetic composition, and this can result in totally different numerical dynamics (fig. 8.8c,d). If there are now subsequent migration events, they may alter the genetic composition of the original founding populations, and this will have a corresponding profound effect on subsequent dynamics (fig. 8.8e,f). In a genetically uniform population, rare migration events subsequent to an initial colonization would have almost no impact.

In natural systems, additional forces may come into play. For example, the level of resistance may be altered by the colonization process itself. Thus Alexander (1989) found that resistant plants produce fewer flowers than susceptible plants; susceptible individuals may therefore be over-

TABLE 8.4

Difference equation models describing local dynamics of numbers of hosts and pathogens for different levels of genetic variation in host resistance and pathogen virulence. Equations assume frequency-dependent disease transmission and density-dependent growth of healthy but not diseased class.

a. Both host and pathogen invariant

$$X_{t+1} = X_t[1 + r - (\beta Y_t)/N_t]$$

$$Y_{t+1} = Y_t[1 + (\beta X_t)/N_t - d]$$

b. Host variation in resistance; pathogen invariant

$$X_{1,t+1} = X_{1,t}[1 + r_1 - (\beta_1 Y_t)/N_t]$$

$$X_{2,t+1} = X_{2,t}[1 + r_2 - (\beta_2 Y_t)/N_t]$$

$$Y_{t+1} = Y_t[1 + (1/N_t)(\beta_1 X_{1,t} + \beta_2 X_{2,t}) - d]$$

c. Host invariant; pathogen variation for virulence

$$X_{t+1} = X_t[1 + r - (1/N_t)(\beta_1 Y_{1,t} + \beta_2 Y_{2,t})]$$

$$Y_{1,t+1} = Y_{1,t}[1 + (\beta_1 X_t)/N_t - d_1]$$

$$Y_{2,t+1} = Y_{2,t}[1 + (\beta_2 X_t)/N_t - d_2]$$

d. Host variation for resistance; pathogen variation for virulence

$$X_{1,t+1} = X_{1,t}[1 + r_1 - (1/N_t)(\beta_{11} Y_{1,t} + \beta_{12} Y_{2,t})]$$

$$X_{2,t+1} = X_{2,t}[1 + r_2 - (1/N_t)(\beta_{21} Y_{1,t} + \beta_{22} Y_{2,t})]$$

$$Y_{1,t+1} = Y_{1,t}[1 + (1/N_t)(\beta_{11} X_{1,t} + \beta_{21} X_{2,t}) - d_1]$$

$$Y_{2,t+1} = Y_{2,t}[1 + (1/N_t)(\beta_{12} X_{1,t} + \beta_{22} X_{2,t}) - d_2]$$

Explanation of symbols: $X_{i,t}$ and $Y_{i,t}$ are respectively the numbers of the ith host and pathogen type at time t; N_t is the total population $(X_t + Y_t)$ at time t. The host growth rate r_i is given by the hyperbolic decay function: $b_{0,i}/(b_{1,i}N_t + 1) - d_i$, where $b_{0,i}$ is the maximum per capita reproduction of the ith host type (at $N_t = 0$), $b_{1,i}$ is a constant that determines the strength of the density dependence, and d_i is a constant death rate for the ith host (or pathogen) type. Disease is transmitted at a rate $\beta_{ij} X_{i,t} Y_{i,t}/Nt$ (i.e., proportional to the absolute density of healthy and the frequency of infective individuals), where β_{ij} is the transmission parameter for the ith host and jth pathogen.

represented in the migrant pool. It is well known from the agricultural literature that plants may show genetic variation for extremes of resistance and susceptibility to pathogens, which in turn may show variation in virulence and avirulence. Such resistance and virulence types may often show gene-for-gene interactions, where susceptibility of a particular host genotype to the disease is contingent on the presence of a particular virulence gene in the pathogen (Flor 1956; Barrett 1985; Burdon 1987). Although relatively little work has been done on natural populations, there is already strong evidence that many natural populations are highly het-

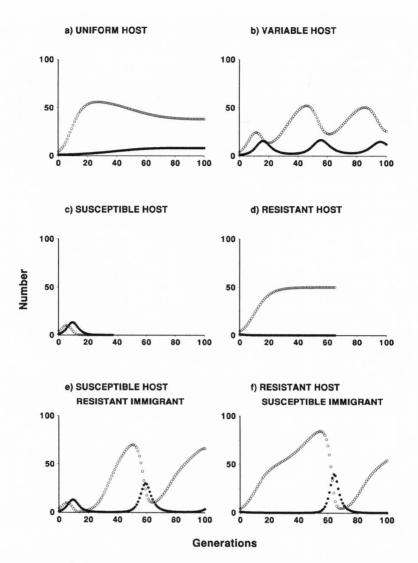

FIG. 8.8. Results from computer simulations showing how the numerical dynamics of small, newly established populations are influenced by genetic composition of the source population and of subsequent immigrants. Recursions on which the models are based are given in table 8.4b. Throughout, $d = 0.3$, $b1 = 0.02$. All populations are started with four healthy (open circles) and one diseased (solid circles) individual. (a) Uniform population of average resistance (with parameter values that result in the same numbers of healthy and diseased individuals at equilibrium as the variable population). (b) Variable population started with two susceptible ($\beta_1 = 1.0$, $b0_1 = 0.9$), and two resistant ($\beta_2 = 0.1$, $b0_2 = 0.6$) individuals. (c) Uniform population started with all susceptible individuals. (d) Uniform population started with all resistant individuals. (e) An initially uniform susceptible population into which there is, at generation 15, a single resistant migrant. (f) An initially resistant population into which there is, at generation 15, a single susceptible migrant.

erogeneous in their resistance/virulence structure. In the *Amphicarpaea bracteata–Synchytrium decipiens* system (Parker 1985), pathogen isolates were virulent on plants from the site where the pathogen was collected but largely avirulent on plants from other sites. In the *Linum marginale–Melampsora lini* system (Jarosz and Burdon 1991), local semi-isolated populations were extremely heterogeneous in their resistance and virulence structure, with no clear relationship between the virulence of particular fungal isolates and the resistance of the host plants from those same populations. However, on a broader geographical scale, all host-resistance types were matched by a corresponding pathogen-virulence type. Whether or not there is local host specialization may depend on the interconnectedness among the component populations.

We might expect that in a metapopulation context, increasing genetic complexity (measured in terms of the number of "matching" resistance/virulence alleles) could lead to less overall disease incidence. Disease spread would require that a pathogen virulence genotype finds a "matching" susceptible host, something that may be particularly difficult if there is a high rate of population extinction/colonization. If this expectation is borne out, the genetic complexity of host-pathogen systems may provide an explanation for the low disease prevalence seen in many natural host-pathogen systems. Somewhat paradoxically therefore, host-pathogen co-evolution may lead to lower disease levels at the metapopulation level.

From a genetic perspective, it has also been shown that complex genetic determination can lead to severe host-pathogen cycles within populations (Seger 1988; see chapter 7, this volume) such that chance extinction of genetic variants is a likely possibility. However, such cycles may be stabilized by migration among populations (Seger 1988; Frank 1991). A metapopulation structure may therefore be important in maintaining genetic variation. Using a computer simulation, Frank (1991) concluded that spatial substructuring of coevolutionary systems can maintain high levels of polymorphism among environmentally identical patches even with high migration rates.

Our eventual goal is to explicitly incorporate genetic variation in host resistance and pathogen virulence into metapopulation models and to ask how it affects the numerical dynamics. We will then be able to ask whether, for example, there are explanations for some of the unusual aspects of our long-term census data. For example, our census data have produced consistent "incidence patterns": small populations are less likely to have disease, but when they are diseased, they have a higher proportion of disease than large populations. We can ask what combination of within- and among-population processes produce such patterns. Specific incidence functions have been predicted for single-species metapopulations (Taylor, B. 1991), but we do not know what to expect in coevolutionary systems. Patterns emerging from our models that are reflected in the empirical data may therefore suggest processes that are important in nature. The simulations will also enable us to ask if some critical initial disease frequency (or genotype frequency) is necessary before a disease is likely to spread. The extreme patchiness in pathogen distribution may result from single colonists being ineffective at establishing the disease. Instead there could be a wave of advance from adjacent populations that might initially have achieved high disease levels purely by chance. The real-world importance of metapopulation processes in disease dynamics has been emphasized recently in a model of the spread of AIDS in situations where populations are organized into villages (May and Anderson 1990). An increase in the interconnectedness of village populations resulted in an initial short-term decrease in overall disease incidence, followed by a subsequent shift of the disease from initially low equilibrium endemic levels to epidemic status.

Conclusions

Much of metapopulation dynamics theory has assumed that within-population dynamics is rapid relative to rates of population extinction and colonization, and that the dynamics of these systems can be represented by presence or absence in patches characterized by singular colonization and extinction probabilities. Our results show that population extinction and colonization rates in the *Silene-Ustilago* system are very rapid, whereas theoretical models indicate that approach to equilibrium within populations may be quite slow (Alexander and Antonovics 1988; Thrall, Biere, and Uyenoyama 1993; and chapter 7, this volume). In such circumstances, which may be quite general for host-pathogen and other coevolutionary systems, it is necessary to understand how within-population processes impact on metapopulation behavior. The issue becomes even more critical when, as in the example presented here, the trajectories of the within population dynamics may be greatly affected by the genetic composition of the initial colonizers and the subsequent migrants. We clearly have to entertain the idea that population interconnectedness may have a large effect on local and metapopulation dynamics and that local populations may be rarely in equilibrium, either genetically or ecologically. However, we hope this study indicates that the expectation of local "nonequilibrium" is not a cause for pessimism, but instead presents a fascinating challenge for an ecological geneticist willing to drive around the countryside.

Acknowledgments

We wish to thank François Felber, Helen Alexander, Arlan Maltby, and Kathy Lemon for initiating the population census amid doubts, uncertainties, and accusations of crass descriptionism. Many people over the years have helped with the census, among them Helen Young, Bernie Roche, Lorraine Kohorn, Julie Carlin, Kaius Helenurm, Chris Richards, and Sheila St. Amour. We wish to thank them for their willingness to suffer dawn awakenings, threatening dogs, poison ivy, and chewy granola bars for breakfast.

References

Alexander, H. M. 1989. An experimental field study of anther-smut disease of *Silene alba* caused by *Ustilago violacea*: Genotypic variation and disease incidence. *Evolution* 43: 835–847.

Alexander, H. M. 1990a. Epidemiology of anther-smut infection of *Silene alba* caused by *Ustilago violacea*: Patterns of spore deposition and disease incidence. *J. Ecol.* 78: 166–179.

Alexander, H. M. 1990b. Dynamics of plant/pathogen interactions in natural plant communities. In *Pests, Pathogens, and Plant Communities*, ed. J. J. Burdon and S. R. Leather, pp. 31–45. Blackwell, Oxford.

Alexander, H. M., and J. Antonovics. 1988. Disease spread and population dynamics of anther-smut infection of *Silene alba* caused by the fungus *Ustilago violacea*. *J. Ecol.* 76: 91–104.

Alexander, H. M., J. Antonovics, and A. W. Kelly. 1993. Genotypic variation in plant disease resistance: Physiological resistance in relation to field disease transmission. *J. Ecol.* 81: 325–333.

Antonovics, J. 1992. Towards community genetics. In *Plant Resistance to Herbivores and Pathogens: Ecology, Evolution and Genetics*, ed. R. S. Fritz and E. L. Simms, pp. 426–429. University of Chicago Press, Chicago.

Antonovics, J., and H. M. Alexander. 1992. Epidemiology of anther-smut infection of *Silene alba* caused by *Ustilago violacea*: Patterns of spore deposition in experimental populations. *Proc. Roy. Soc. London B* 250: 157–163.

Antonovics, J., and D. A. Levin. 1980. The ecological and genetic consequences of density-dependent regulation in plants. *Ann. Rev. Ecol. Syst.* 11: 411–452.

Barrett, J. A. 1985. The gene-for-gene hypothesis: Parable or paradigm. In *Ecology and Genetics of Host-Parasite Interactions,* ed. D. Rollinson and R. M. Anderson, pp. 215–225. Academic Press, London.

Brown, J. H., and A. Kodric-Brown. 1977. Turnover rates in insular biogeography: Effect of immigration on extinction. *Ecology* 58: 445–449.

Burdon, J. J. 1987. *Disease and Plant Population Biology.* Cambridge University Press, Cambridge, U.K.

Burkey, T. V. 1989. Extinction in nature reserves: The effect of fragmentation and the importance of migration between reserve fragments. *Oikos* 55: 75–81.

Caswell, H. 1978. Predator-mediated coexistence: A non-equilibrium model. *Amer. Nat.* 112: 127–154.

Diekmann, O., J.A.J. Metz, and M. W. Sabelis. 1988. Mathematical models of predator-prey-plant interactions in a patchy environment. *Exp. and Appl. Acarology* 5: 319–342.

Flor, A. H. 1956. The complementary genic system in flax and flax rust. *Adv. in Genet.* 8: 29–54.

Frank, S. A. 1991. Spatial variation in coevolutionary dynamics. *Evol. Ecol.* 5: 193–217.

Getz, W. M., and J. Pickering. 1983. Epidemic models: Thresholds and population regulation. *Amer. Nat.* 121: 892–898.

Gilpin, M. E. 1975. *Group Selection in Predator-Prey Communities*. Princeton University Press, Princeton, N.J.

Gilpin, M. 1991. The genetic effective size of a metapopulation. *Biol. J. Linnean Soc.* 42: 165–175.

Gyllenberg, M., and I. Hanski. 1992. Single species metapopulation dynamics: A structured model. *Theor. Pop. Biol.* 42: 35–61.

Hanski, I. 1982. Dynamics of regional distribution: The core and satellite species hypothesis. *Oikos* 38: 210–221.

Hanski, I. 1991. Single-species metapopulation dynamics: Concepts, models and observations. *Biol. J. Linnean Soc.* 42: 17–38.

Hanski, I., and M. Gilpin. 1991. Metapopulation dynamics: Brief history and conceptual domain. *Biol. J. Linnean Soc.* 42: 3–16.

Harrison, S. 1991. Local extinction in a metapopulation context: An empirical evaluation. *Biol. J. Linnean Soc.* 42: 73–88.

Hastings, A., and C. L. Wolin. 1989. Within-patch dynamics in a metapopulation. *Ecology* 70: 1261–1266.

Huffaker, C. B. 1958. Experimental studies on predation: Dispersion factors and predator-prey oscillations. *Hilgardia* 27: 343–383.

Hutchinson, G. E. 1953. The concept of pattern in ecology. *Proc. Natl. Acad. Sci. USA* 105: 1–12.

Jarosz, A. M., and J. J. Burdon. 1991. Host-pathogen interactions in natural populations of *Linum marginale* and *Melampsora lini*: II. Local and regional variation in patterns of resistance and racial structure. *Evolution* 45: 1618–1627.

Jennersten, O., S. G. Nilssons, and U. Wastljung. 1983. Local plant populations as ecological islands: The infections of *Viscaria vulgaris* by the fungus *Ustilago violacea*. *Oikos* 41: 391–395.

Kareiva, P. 1984. Predator-prey dynamics in spatially structured populations: Manipulating dispersal in an coccinelid-aphid interaction. *Lecture Notes in Biomath.* 54: 368–389.

Kareiva, P. 1987. Habitat fragmentation and the stability of predator-prey interactions. *Nature* 326: 388–390.

Levins, R. 1969. Some demographic and genetic consequences of environmental heterogeneity for biological control. *Bull. Entomol. Soc. America* 15: 237–240.

MacArthur, R. H., and E. O. Wilson. 1967. *The Theory of Island Biogeography.* Princeton University Press, Princeton, N.J.

McCauley, D. E. 1991. Genetic consequences of local population extinction and recolonization. *Trends in Ecol. and Evol.* 6: 5–8.

McNeill, J. 1977. The biology of Canadian weeds. 25. *Silene alba* (Miller) E.H.L. Krause. *Canadian J. Plant Sci.* 57: 1103–1114.

Martin, G. B., J.G.K. Williams, and S. D. Tanksley. 1991. Rapid identification of markers linked to a *Pseudomonas* resistance gene in tomato by using random primers and near isogenic lines. *Proc. Natl. Acad. Sci. USA* 88: 2336–2340.

May, R. M. 1978. Host-parasitoid systems in patchy environments: A phenomenological model. *J. Animal Ecol.* 43: 747–770.

May, R. M., and R. M. Anderson. 1990. Parasite-host coevolution. *Parasitology* 100 (suppl.): 89–101.

Murdoch, W. W. 1977. Stabilizing effects of spatial heterogeneity in predator prey systems. *Theoret. Pop. Biol.* 11: 252–273.

Nisbet, R. M., and W.S.C. Gurney. 1982. *Modelling Fluctuating Populations.* Wiley, New York.

Parker, M. A. 1985. Local population differentiation for compatibility in an annual legume and its host-specific pathogen. *Evolution* 39: 713–723.

Peltonen, A., and I. Hanski. 1991. Patterns of island occupancy explained by colonization and extinction rates in shrews. *Ecology* 72: 1698–1708.

Pimentel, D., W. P. Nagel, and J. L. Madden. 1963. Space-time structure of the environment and the survival of parasite-host systems. *Amer. Nat.* 97: 141–166.

Rey, J. R., and D. R. Strong. 1983. Immigration and extinction of salt-marsh arthropods on islands: An experimental study. *Oikos* 41: 396–401.

Sabelis, M. W., O. Diekmann, and V.A.A. Jansen. 1991. Metapopulation persistence despite local extinction: Predator-prey patch models of the Lotka-Volterra type. *Biol. J. Linnean Soc.* 42: 267–283.

Sabelis, M. W., and W.E.M. Laane. 1986. Regional dynamics of spider-mite populations that become extinct locally because of food source depletion and predation by phytoseiid mites (Acarina: Tetranychidae, Phytoseiidae). In *Dynamics of Physiologically Structured Populations*, ed. J.A.J. Metz and O. Diekmann, p. 345. Lecture Notes in Biomathematics 68. Springer-Verlag, Berlin.

Sabelis, M. W., and J. van der Meer. 1986. Local dynamics of the interaction between predatory mites and two spotted spider mites. In *Dynamics of Physiologically Structured Populations*, ed. J.A.J. Metz and O. Diekmann, pp. 322–344. Lecture Notes in Biomathematics 68. Springer-Verlag, Berlin.

Seger, J. 1988. Dynamics of some simple host-parasite models with more than two genotypes in each species. *Philos. Trans. Roy. Soc. London B* 319: 541–555.

Takafuji, A. 1977. The effect of the rate of successful dispersal of a phytoseiid mite, *Phytoseilus persimilis* (Acarina: Phytoseiidae) on the persistence in the interactive system between the predator and its prey. *Res. Pop. Ecol.* 18: 210–222.

Taylor, A. D. 1991. Studying metapopulation effects in predator-prey systems. *Biol. J. Linnean Soc.* 42: 305–323.

Taylor, B. 1991. Investigating species incidence over habitat fragments of different areas—a look at error estimation. *Biol. J. Linnean Soc.* 42: 177–191.

Thrall, P. H. 1993. Population and genetic dynamics of the Anther-Smut *Ustilago violacea* and its plant host *Silene alba*. Ph.D. thesis, Duke University, Durham, N.C.

Thrall, P. H., J. Antonovics, and D. W. Hall. 1993. Host and pathogen coexistence in vector-borne and venereal diseases characterized by frequency dependent disease transmission. *Amer. Nat.* 142: 543–552.

Thrall, P. H., A. Biere, and M. Uyenoyama. 1993. The population dynamics of host-pathogen systems with frequency dependent disease transmission. *Amer. Nat.*, in press.

Uyenoyama, M., and M. W. Feldman. 1980. Theories of kin and group selection: A population genetics perspective. *Theor. Pop. Biol.* 17: 380–414.

Verboom, J., and K. Lankester. 1991. Linking local and regional dynamics in stochastic metapopulation models. *Biol. J. Linnean Soc.* 42: 39–55.

Wilcox, B. A. 1990. In situ conservation of genetic resources. In *The Preservation and Valuation of Biological Resources*, ed. G. H. Orians, G. M. Brown, W. E. Kunin, and J. E. Swierzbinski, pp. 45–77. University of Washington Press, Seattle.

Williams, J.G.K., A. R. Kubelik, K. J. Livak, J. A. Rafalski, and S. V. Tingey. 1990. DNA polymorphisms amplified by arbitrary primers are useful as genetic markers. *Nucleic Acids Res.* 18: 6531–6535.

Wilson, D. S. 1983. The group selection controversy: History and current status. *Ann. Rev. Ecol. and Syst.* 14: 159–187.

Wright, S. 1943. Isolation by distance. *Genetics* 28: 114–138.

9

Ecological Genetics of Life-History Traits: Variation and Its Evolutionary Significance

Introduction

A major goal of evolutionary biology has been the development of a general theory that will allow the accurate prediction of those life-history traits most likely to evolve in different ecological settings. This goal is motivated by two sets of observations. First, within any well-defined phylogenetic group, life-history traits vary nonrandomly with respect to ecological setting (Cody 1966; Pianka 1970; Baker 1972; Brewer and Swander 1977; May and Rubenstein 1985; Primack 1985; Dunham et al. 1988; Wilbur and Morin 1988). For example, actively foraging lizard species have smaller clutch sizes than sit-and-wait foragers (Dunham et al. 1988), and species of buttercup (*Ranunculus*) that occur in drier habitats have larger seeds (Baker 1972). Second, specific patterns of covariation among life-history traits, like the inverse relation between offspring size and number, recur in many groups (Primack 1987; Rohwer 1988; Elgar and Heaphy 1989; Mazer 1989; Mitton and Lewis 1989; Read and Harvey 1989; Reznick and Miles 1989; Elgar 1990), which suggests that there must be general rules for the organization of life-history variation.

Mathematical models dedicated to understanding the genesis and maintenance of these patterns examine how different rates of age- or stage-specific mortality select for different optimal schedules of reproductive timing, investment, and packaging. The problem is made tractable by adding a set of a priori constraints on the permissible schedules. This approach to the problem is motivated by another common pattern that emerges from broad taxonomic surveys: some measure of average stage-specific mortality rate is always found to be correlated with the average value of a life-history trait like clutch size or the age at first reproduction (Millar and Zammuto 1983; Saether 1988; Charnov 1991). Any particular model can be connected to a particular association of life-history variation and ecological variation (e.g., active vs. sit-and-wait foraging in lizards or moisture regime for buttercups) by examining how the ecological distinctions lead to distinctions in mortality patterns that may select for different life-history patterns (MacArthur and Wilson 1967, chap. 7). The relative importance of specific sources of mortality or exactly which environmental factors constrain permissible trait values, for example whether resource limitation or predation is the more common agent of mortality, is an empirical ecological problem that is independent of the theoretical linkage between specific mortality regimes and constraints and specific ensembles of life-history traits (Wilbur et al. 1974; Boyce 1984).

The diverse ecological contexts for life-history variation have inspired a diversity of specific models. Some models address only which phenotypes are optimal for a given demographic milieu and set of constraints (Goodman 1982; Schaffer 1983; McGinley et al. 1987; Winkler and Wallin 1987; Pugliese and Kozlowski 1990), while others ask if a genetic model of life-history variation can achieve the optimal phenotype (Charlesworth 1980; Lande 1982; Rose 1982, 1985). Some models apply only to density-independent population dynamics (Law 1979), others include density-dependence (Michod 1979), and others are concerned explicitly with the distinction between unregulated and regulated populations (MacArthur and Wilson 1967). Other distinctions exist in the postulated constraints and the details of model construction and analysis (Taylor et al. 1974; Iwasa and Teramoto 1980; McGinley et al. 1987; Perrin 1992).

Mathematical life-history theory has been remarkably successful. Simple, explicit predictions about reproductive timing and investment have been confirmed by laboratory selection experiments (Luckinbill 1978, 1984; Luckinbill et al. 1984; Luckinbill and Clare 1986; Rose and Charlesworth 1981; Mueller 1988; Spitze 1991). Several paradoxical laboratory results are traceable to either a failure in experimental protocol or a failure to generate the correct prediction for the relevant mode of population dynamics (see discussions in Bergmans 1984; Mueller 1988; Perrin 1988). Several experimental field studies have demonstrated the connection between divergent ecological circumstances and divergent mortality patterns and shown that the general predictions of life-history theory are upheld in genetically based trait variations (Law et al. 1977; Schemske 1984; Reznick et al. 1990).

These advances confirm that broad differences in the average schedules of reproductive timing and investment evolve in response to broad differences in average age- and stage-specific mortality rates. However, a significant amount of natural life-history diversity cannot be explained solely in this context; the evolutionary linkage of mortality rate and investment rate is a necessary but insufficient principle for understanding the sources of natural life-history diversity. This claim is the first thesis of this review. The reasons for this claim and a hypothesis for which principles must be added are, taken together, its second thesis.

The central argument of this review is based on the fact that any observed "life history" is in fact the phenotypic manifestation of physiological processes that are readily susceptible to environmental influences. I will argue that the specific interaction of genotype and environment that produces a sensitivity of trait expression to environmental influences is as much the product of life-history evolution as are average values of individual life-history traits. This is the principle that must be integrated into future work if we are to fully uncover the sources of natural life-history diversity. The challenge for the next generation of theoretical and empirical work is to understand how these interactions are molded by natural selection through age- and stage-specific mortality rates in a variable environment to express an optimal *set* of life histories for the distribution of environmental conditions that is encountered.

In this chapter I first review interspecific variation in life history to illustrate the phenomena that have inspired the entire field of inquiry and to show why the focus of empirical investigation was turned onto intraspecific variation. In the succeeding section I review studies of intraspecific variation and illustrate what these studies have revealed about the causes of life-history diversity. In the third section I review the experimental evidence that phenotypic plasticity is a necessary element of a full explanation of life-history diversity and not merely a nuisance that obscures our ability to test modern theory. In the final section I offer a connection between life-history theory and the study of plasticity that outlines the directions along which further work might profitably proceed.

Interspecific Variation in Life History

The extent of variation in life history within almost any well-defined group of organisms is astonishing. For example, within the poeciliid fishes, a monophyletic subfamily of cyprinodontiform fishes (Parenti and Rauchenberger 1989), average brood size and mass of individual offspring vary among species by an order of magnitude, even after adjustment for interspecific differences in female body size (Reznick and Miles 1989). Even within the genus *Poecilia* the average brood size varies almost sixfold, again even after adjustment for interspecific differences in female body size (Reznick and Miles 1989). Comparable variation exists in many other groups. Among genera of iguanid lizards, average age at maturity and clutch size vary almost ninefold (Miles and Dunham 1992). Species of *Plantago* exhibit a fourfold range in the average number of seeds per unit leaf area (Primack 1979).

A striking aspect of this broad-scale variation is that specific patterns of covariation among traits are repeated in almost every group. The compromise between offspring size and offspring number is only the most appreciated of several such correlations. For example, semelparous mollusks devote almost twice the biomass to each reproductive episode as do iteroparous mollusks (Browne and Russell-Hunter 1978), a pattern reflected in the contrast between annual and perennial species of the cosmopolitan plant genus *Plantago* (Primack 1979). Delayed maturity is repeatedly associated with lower rates of reproductive effort at each reproductive bout (Cody 1971; Dunham et al. 1988). These repeated covariances could reflect either general results of repeatable patterns of selection or general sets of constraints on how life histories can evolve.

The actual level of interspecific variation within a group may not reflect the amount of evolutionary radiation in life histories that has occurred. Extensive cladogenesis might have occurred with only little modification of the life histories of daughter taxa. Several approaches have been taken to discern the hierarchical level at which most life-history diversification has occurred (Cheverud et al. 1985; Felsenstein 1988; Pagel and Harvey 1988; Bell 1989; Burt 1989; Grafen 1989; Brooks and McLennan 1991). As one might expect, different groups exhibit different hierarchical patterns. Within eutherian mammals, the lion's share of variation in most traits occurs at the ordinal level (Read and Harvey 1989). On the other hand, variation among lizards in clutch size appears to have been molded by species-specific evolution that is independent of phylogenetic affinity (Dunham et al. 1988; Miles and Dunham 1992).

Phylogenetic considerations introduce a major dilemma for employing broad taxonomic surveys to understand the adaptive evolution of life history because of the inevitable correlation of phylogenetic heritage and ecological habit. For example, Kaplan and Salthe (1979) showed that salamander species that bred in ponds had smaller ovum volumes, even when adjusted for variation in female body size, than salamander species that bred in streams, which in turn had smaller ovum volumes than species with direct development. However, breeding habitat of salamanders is highly associated with family identity. A similar confounding occurs with the incidence of multiple clutching within a single season in temperate birds. Forest birds are single brooded, while birds in more open habitats often produce multiple broods; however, these patterns are mainly due to differences in the distribution of families among habitats because species within a family are consistently either singly or multiply brooded (Brewer and Swander 1977).

Such confounding makes the patterns consistent with several hypotheses. The most tempting hypothesis is that some feature of breeding habitat selects consistently for the observed life-history differences. Evolutionary responses might occur in two ways, however. First, each species in each habitat may have undergone independent adaptive modification of the trait in question. Alternatively, it is possible that families are constrained in ovum size (salamanders) or multiple clutching (birds) and that such constraints forced their constituent species to occupy only specific habitats. Adaptive evolution of trait values has occurred in each case, but at one extreme (the former explanation) evolution of trait values occurs through direct selection on the trait while at the other extreme (latter hypothesis) differential extinction of taxa with inappropriate trait values generates the pattern. One can even postulate a more radical scenario: there is no adaptive significance to the differences in ovum size or clutch number at all, but they are ineluctable correlates of selection on some other aspect of the organism and have been constrained to respond through indirect selection.

One way to extricate the study of life histories from this dilemma is to demonstrate that the same association of trait values and ecological conditions occurs repeatedly among sets of species drawn from different families or genera. Baker (1972) and Mazer (1989) have shown that patterns of seed size variation among species are repeated across many such sets of taxa in California and Indiana floras, respectively. However, in some groups it may be almost impossible to disentangle these confounded factors (e.g., Brown 1983).

Another approach has been to perform comparative ecological studies on a circumscribed set of related species that differ in life histories. Several such studies have shown a congruence between interspecific differences in mortality schedules and differences in life histories and have thereby been offered as support for one or more individual models of life-history evolution (Gadgil and Solbrig 1972; Ballinger 1973; Gaines et al. 1974; Wilbur 1976; Brown 1979). Yet even these studies can be viewed as problematic with respect to their usefulness as decisive tests of theory (Stearns 1976, 1977). They often confound differences in ecology with differences in phylogeny, and they inevitably confound differences in demographic environment with differences in ontogenetic conditions that might influence trait expression. They also cannot illuminate the genetic basis of differences in individual traits or a suite of correlated traits. The lack of definitive genetic information imposes two limitations on the possible inferences. First, it is impossible to separate the magnitude of genetic differences in life-history trait values from environmental effects on those values. The ease with which such environmental effects can be detected (fig. 9.1) imparts a premium on this separation, without which it is impossible to discern just how finely selection can match life-history differences to demographic differences. Second, it is impossible to interpret differences in several traits as the result of direct selection on each trait.

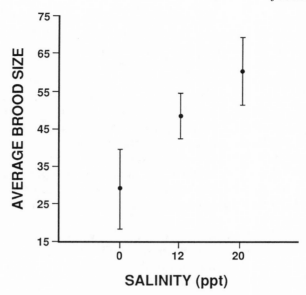

FIG. 9.1. Average brood size in *Poecilia formosa* at 29°C as a function of salinity, expressed in parts per thousand. Each point represents the average value for 12 females, and the bars indicate one standard error above and below the average. (Figure based on unpublished data of the author, J. C. Trexler, and C. D. Johnson)

An unknowable proportion of trait differences might have arisen from indirect selection and the correlated responses produced by close linkage or pleiotropy. Without some knowledge of genetics as well as the nature of selection, it is impossible to discern whether covarying trait differences represent finely honed adaptation or the inevitable result of constraints on the permissible trait combinations that channel the evolutionary response to demographic selection.

Ironically, although broad interspecific comparisons motivated the entire field of study, these deductions cast doubt on the extent to which such comparisons can illuminate the correct mechanisms of evolutionary diversification. One response to this dilemma is to recast the question at the population level so that genetic analyses could be made.

Variation among Conspecific Populations

Conspecific populations often differ widely in their average values of individual life-history traits. Average age at maturity varies from one to seven years among populations of the European minnow, *Phoxinus phoxinus*, and size-specific fecundity varies fourfold (Mills 1988). Even endothermic animals can show appreciable variation: age at maturity varies twofold among local populations of the ground squirrel *Spermophilus columbianus* (Zammuto and Millar 1985a). Size-specific fecundity varies markedly among populations of many poeciliid fish; populations of Hawaiian mosquitofish vary almost ninefold between extremes (Stearns 1983a), sailfin mollies exhibit a nearly eightfold range (Travis 1989), and populations of guppies vary regularly across a twofold range (Reznick and Endler 1982; Reznick 1989). Average seed mass among conspecific plant populations can vary two- to fourfold (Wulff 1986a; Winn and Werner 1987).

Not only do average phenotypic values of life-history traits vary among populations but the physiological mechanisms through which those values are attained may vary. One example is the variation in maternal provisioning seen among populations of the sailfin molly (Trexler 1985). Poeciliid species have been known to vary in whether females provide any nutrition to embryos after fertilization; lecithotrophic species do not, but matrotrophic species do so via either a direct vascular connection between female and embryo or contact between maternal bloodstream and absorptive cells on the embryo's surface (Wourms et al. 1988). Trexler showed that some populations of mollies were lecithotrophic while others in the same geographic area were matrotrophic, and that variation among populations in neonate mass could be generated by this variation in the mode of maternal provisioning. Wulff (1986b) showed that the wide variation in seed size in *Desmodium paniculatum* was produced by variation in the extent of maternal provisioning and not by intrinsic differences in embryo size.

In many cases the patterns of intraspecific variation matched the patterns of interspecific variation. For example, seed size variation among populations of *Prunella vulgaris* reflected interspecific associations of larger size with decreased light availability and disturbance frequency (Winn and Werner 1987). The common interspecific associations of larger seeds with drier environments (Baker 1972; Mazer 1989) were found to be repeated intraspecifically among populations of *Amaranthus retroflexus* (McWilliams et al. 1968; Schimpf 1977). Patterns of life-history variation among populations of *Plantago lanceolata* reflected the differences between "successional" and "climax" species that were often attributed to r- and K-selection (Antonovics and Primack 1982). Intraspecific variation in Hawaiian mosquitofish showed the same structure as that in Trinidadian guppies (Reznick and Miles 1989); this structure reflected the "trade-off" between offspring size and number seen among species of poeciliid fish after adjustment for body-size variation.

In some cases the ecological associations of intraspecific variation in a trait have not been consistent with those found in interspecific compari-

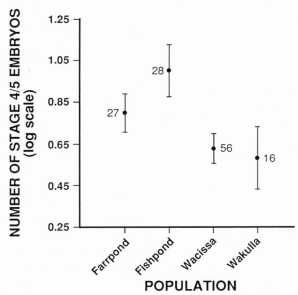

FIG. 9.2. Average number of offspring per brood (logarithmic scale) for female *Heter-andria formosa* collected from four natural populations in north Florida. Bars represent one standard error above and below the average. Farrpond and Fishpond are small lakes; Wacissa and Wakulla are rivers. (Figure taken from data in Forster-Blouin 1989)

sons. The patterns of variation in life-history traits and demography among populations of two lizard species, *Urosaurus ornatus* and *Scelo-porus undulatus*, were in no way comparable to the patterns of variation seen among species and could not be interpreted in light of any mathematical theory (Dunham 1982; Tinkle and Dunham 1986).

Intraspecific variation in life-history traits is rarely at random with respect to ecological factors. For example, brood sizes in the poeciliid fish *Heterandria formosa* are higher in populations that inhabit still water than in those in running water (fig. 9.2), a pattern that reflects interspecific variation seen within the tribe Heterandriini (Constantz 1979). A diversity of workers have found intraspecific life-history variation to be correlated with variation in ecological setting and, with varying degrees of surety, argued that such differences have a genetic basis, are adaptive, and reflect accurate theoretical predictions (Tinkle and Ballinger 1972; Schaffer and Elson 1975; Law et al. 1977; Leggett and Carscadden 1978; Constantz 1979; Dingle 1981; Dingle et al. 1980, 1982; Berven 1982a,b; Stearns 1983a,b; Reinartz 1984a,b,c; Schemske 1984; Semlitsch and Gibbons 1985; Zammuto and Millar 1985a,b; Wyngaard 1986a; Etter 1989;

Reznick 1982a,b, 1989; Reznick and Endler 1982; Reznick et al. 1990). In the most convincing of these studies, workers examined several replicate populations from each type of setting and could actually demonstrate experimentally the genetic basis of the relevant life-history variation and its adaptive significance.

Perhaps the most surprising and provocative result of many intraspecific studies has been the discovery that differences among populations that appear to be in the "correct" direction for an adaptive interpretation have little to no genetic basis. Brown (1985) found that the variation in every trait among four populations of the pond snail *Lymnaea elodes* was entirely due to phenotypic plasticity. Winn's (1985, 1988) and Winn and Werner's (1987) study of seed mass variation in *Prunella vulgaris* is an especially notable example. Patterns of seed size variation among populations reflected the interspecific correlation of larger seeds with more shaded environments. However, the differences among populations were almost entirely environmental in origin, and there were no differences among the habitats in the effects of seed size on seedling establishment. Other notable examples of striking life-history plasticity include Fowler and Antonovics (1981) and Platenkamp (1990, 1991).

Even when population differences appear to have a genetic basis, the magnitude and direction of that genetic distinction is usually dependent upon the level of some environmental factor (a result known at least since Vetukhiv 1957). This phenomenon can cloud the interpretation of many life-history studies. One striking example is the work of Primack and Antonovics (1982) with *Plantago lanceolata*; although phenotypic variation in life history and demographic variation among populations matched theoretical predictions nearly perfectly, the differences among populations that were observed in a common-garden experiment bore little relation to the phenotypic patterns seen in the natural populations.

Schmidt and Levin's (1985) study of population differentiation in *Phlox drummondii* is even more striking. They performed reciprocal transplants among eight populations in each of two years. The effect of location of study accounted for about 40% of the phenotypic variation in fecundity in each year. In each year there was a significant effect of population of origin, which accounted for 14% of the phenotypic variance in fecundity in one year and 4% in the other. However, the plants with different origins actually showed significant differences at only three of the eight transplant locations each year, and only one of those sites was the same in both years. In the first year the pattern of differences among plants from different sources was consistent with the interpretation of local adaptation, but in the second year the pattern of differences was not so interpretable. Clearly, plasticity accounted for most of the natural variation in life history, but the striking site × origin interactions indicated an extensive array of genetic variation in the actual sensitivity and response to environmental effects.

Genetic analyses of covarying trait differences among populations have offered less satisfying results than analyses of individual traits. Studies on poeciliid fishes have indicated some concordance between the apparent constraints inferred from genetic covariances among life-history traits and the patterns of intraspecific divergence for trait combinations (reviewed in Travis 1989). Not all of these examples could withstand critical scrutiny; precise estimates of additive genetic covariances are rare and there are no such estimates of covariances from several natural populations that may be compared to the multivariate pattern of trait variation among those populations. There is some evidence that genetic covariances within milkweed bug populations are in general alignment with the major axes of population differentiation (Palmer and Dingle 1986). More restricted studies on pairs of populations have produced similarly weak results. Bivariate axes of phenotypic differentiation in copepods (Wyngaard 1986a) and frogs (Berven 1982b) are consistent with some estimated covariances, but in each case differences among populations in the magnitudes of genetic variances precluded a thoroughly convincing conclusion because of the problem of statistical power (Wyngaard 1986b; Berven 1987).

The patterns of phenotypic covariances within and among natural populations have proven no more reassuring. Studies of morphological traits have not shown any general concordance between intra- and interpopulation covariances (James, 1991), and neither have studies of life-history covariances (e.g., Wilbur 1977). However, regardless of the statistical patterns of the results, any interpretation of phenotypic or genetic covariances in this context is clouded because in theory the present directions of population covariances need not reflect the initial genetic covariance that directed the response to selection (Turelli 1988; Zeng 1988).

The usefulness of genetic covariances as indicators of evolutionary constraints becomes even more questionable in the face of evidence for substantial environmental effects on genetic expression. Genetic differences in life-history traits between populations of milkweed bugs changed notably in their expression in different rearing environments; traits that showed strong genetic covariances in one environment did not always have parallel changes in expression between environments (Dingle et al. 1982). This result indicates that pleiotropic allelic effects will change across environments, which in turn can alter the genetic covariances among traits from one environment to another and thereby make them unreliable indicators of evolutionary constraints in a variable environment (Clark 1987; Jinks and Pooni 1988; Scheiner and Lyman 1991; Stearns et al. 1991).

These studies on intraspecific variation indicate that phenotypic plasticity in life-history traits is a major cause of the observed diversity among conspecific populations, if not among closely related species. Field observations have long suggested the action of phenotypic plasticity. In many populations life-history traits have been observed to display annual (Vitt et al. 1978; Tinkle et al. 1981; Healey and Dietz 1984; Trexler 1985; Travis and Trexler 1987) and even seasonal (Hubbs et al. 1968; Bagenal 1971; Nussbaum 1981) variation. In some cases seasonal variation within a population embraces a wider range than variation among populations in the same season (Schoenherr 1977; Trendall 1982) and annual variation causes life-history differences among populations to appear in some years but not others (Trexler 1985; this need not be so, cf. Reznick and Bryga 1987).

Controlled experimental studies on the plasticity of life-history traits offer two important lines of evidence for evaluating its potential importance. First, almost every conceivable life-history trait is known to respond to almost every conceivable environmental factor in at least one species of plant or animal. Reciprocal transplant studies have long demonstrated the plasticity of life-history traits (Berven 1982a,b, among many others) and the phenotypic values of life-history traits in plants and animals have been shown to respond experimentally to variation in nutrition level (Hislop et al. 1978; Reznick 1983; Townshend and Wootten 1984; Lacey 1986; Wulff 1986a; Guyer 1988), temperature (Giesel et al. 1982; Murphy et al. 1983; Kaplan 1987), photoperiod (Cook 1975; Giesel 1986; Lacey 1986; Groeters and Dingle 1987), and density (Dahlgren 1979; Rubenstein 1981; Busack and Gall 1983; Smith 1983; Harris 1987; Semlitsch 1987a,b; Mazer and Schick 1991). Phenotypic correlations between individual traits have been altered by experimental treatments, either through manipulations of specific factors (Dahlgren 1980; Murphy et al. 1983) or gross changes in environmental conditions (Primack and Antonovics 1982; Winn and Werner 1987).

Second, and more importantly, the magnitude of the life-history variation that can be induced by environmental variation is often comparable to the magnitude of phenotypic variation seen either among conspecific populations or across seasons or years within a population. Most of the studies cited above support this claim. For example, size-adjusted fecundity in poeciliid fishes usually can be found to vary two- to threefold among populations, a level readily induced by varying food level (Trendall 1983) or other environmental effects (fig. 9.2) within a range that reflects natural environmental variation.

This evidence indicates that phenotypic plasticity can be responsible for a significant amount of natural life-history diversity. However, the significance of phenotypic plasticity goes beyond this role; in the next sections I suggest that plasticity itself is often a vital product of life-history evolution.

Experiments on the Nature of Life-History Plasticity

Several lines of evidence indicate that phenotypic plasticity in life-history traits is a phenomenon of evolutionary importance. First, not all species show the same response of a specific life-history trait to the same type of environmental variation. For example, in many fish an increase in food level produces an increase in clutch or brood size and a decrease in the size of each individual offspring (e.g., Bagenal 1969). In other fish, however, increases in nutrition level increase brood size without affecting neonate size (Wootten 1973). These results show that even the revered "trade-off" between number and size of offspring is malleable (see also Dahlgren 1980).

Interspecific differences in which traits exhibit phenotypic plasticity may be associated nonrandomly with other differences in the life history (Bradshaw 1965; Schlichting 1986). For example, in lecithotrophic poeciliid fishes increases in food level or changes in environment that decrease the maintenance metabolic rate produce increases in brood size but do not alter the rate at which successive broods are produced (Trendall 1983; fig. 9.1). In the most matrotrophic species in the family, increased food level does not increase brood size but does increase the rate at which broods are produced (Travis et al. 1987). Circumstantial evidence suggests that this association will be consistent throughout the poeciliid family but it has not yet been demonstrated (Travis et al. 1987). A remarkably similar pattern of differences in the effects of food level on reproduction is seen in the contrast between an oviparous and a viviparous snake (Ford and Seigel 1989).

Even closely related species can exhibit idiosyncratic patterns of environmental sensitivity. For example, female sailfin mollies (*Poecilia latipinna*) from north Florida show a dramatic increase in juvenile growth rate with increasing temperature and a small to moderate increase with increasing salinity (Trexler et al. 1990). The gynogenetic, all-female derivative of this species, the Amazon molly (*P. formosa*), which shares half of its nuclear genome with the sailfin molly (Avise et al. 1991), shows much less sensitivity of juvenile growth rate to temperature and exhibits a decrease in growth rate with increased salinity (fig. 9.3). These striking differences between the taxa are not seen in other life-history traits. How-

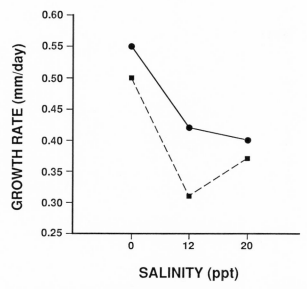

FIG. 9.3. Average growth rate of immature female *P. formosa* as a function of temperature and salinity. Salinity is on the horizontal axis in parts per thousand; squares indicate the averages for 24°C and circles indicate those for 29°C. Each point is the average for 12–14 females at that treatment combination. (Data from same source as fig. 9.1)

ever, the Amazon molly's pattern of plasticity in age and size at maturity with respect to temperature and salinity is identical to that of female sailfin mollies (fig. 9.4). Taken together, these two sets of results indicate that the two taxa have very different developmental pathways that respond qualitatively differently to environmental variation.

The preceding line of evidence indicates that patterns of environmental sensitivity can evolve and may do so nonrandomly with respect to other facets of the life history. The evidence says nothing about whether such patterns evolve through natural selection. However, some patterns of plasticity in individual species appear to have adaptive significance. This is the second line of evidence for the evolutionary significance of plasticity in life-history traits. The induction of diapause responses in many insects by various environmental cues may be the most familiar examples of adaptive plasticity (Tauber et al. 1986). The alternative developmental pathways followed by many cladocerans in response to the presence of predatory midges are also well-known examples (although they are not without problematic features; see review in Spitze 1992). Other examples would include the sensitivity of development patterns in amphibian larvae to the presence of predators (Skelly and Werner 1990) or the onset of

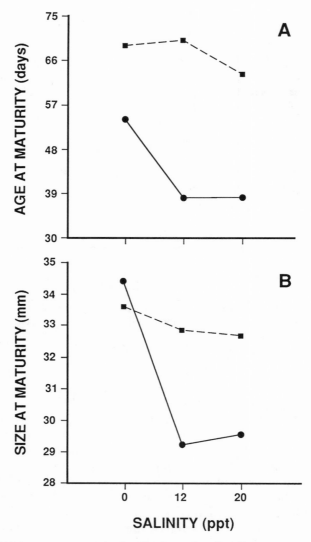

FIG. 9.4. (A) Average age at maturity (days) as a function of temperature and salinity in female *P. formosa*. Symbols and sample sizes are as in figure 9.3. (B) Average standard length (mm) at maturity as a function of temperature and salinity in *P. formosa*. Symbols and sample sizes are as in figure 9.3. (Data from same source as figs. 9.1 and 9.3)

pond drying (Newman 1988) and the dramatic change in the age and size at maturity in the pond snail *Physella virgata* in response to the presence of crayfish predators (Crowl and Covich 1990).

In many cases differences in plasticity among congeneric species or conspecific populations reflect plausible expectations for adaptive differentiation. These expectations are based on the differences among the species or populations in ecological conditions that involve the trait or traits in question and the environmental variation under study (Berven et al. 1979; Dingle et al. 1980; Dingle and Baldwin 1983; Conover and Present 1990; Zangerl and Berenbaum 1990; So and Takafuji 1991). For example, individuals drawn from populations of grasses with an evolutionary history of grazing respond to being grazed by changing their subsequent allocation of resources in favor of increased root biomass; individuals drawn from populations without such a history do not alter their allocation patterns after being grazed (Schimel 1992). These differences in the response to grazing are reasonable expectations for an adaptive divergence to different histories of grazing pressure.

Two case studies of ambystomatid salamanders illustrate this kind of evidence. In north Florida three species breed sequentially in the same habitat from late autumn to late winter; larvae of species that breed later are susceptible to predation from the older, larger larvae of each preceding species. Predation risk vanishes when younger larvae become too large to be eaten by older larvae. The larval growth rates of the species that breed second and third are faster at the same temperatures and are more responsive to temperature variation than those of the species that breeds first (Keen et al. 1984). Later-breeding species appear to have been selected for more rapid growth and greater ability to respond to the seasonal increases in temperature through the enhanced survival that such a pattern of plasticity would bring. Conover and Present (1990) offer a comparable argument for the geographic differentiation in the phenotypic plasticity of growth rate in response to temperature in *Menidia menidia*.

In the second example from the ambystomatid salamanders, the incidence of paedomorphosis varies among populations of *Ambystoma talpoideum* on the Atlantic coastal plain; populations in more permanent ponds are more likely to exhibit paedomorphosis (Semlitsch and Gibbons 1985). Normal metamorphosis can be induced by the drying of the pond. Salamanders from populations that inhabit ponds that dry regularly are more responsive to the environmental changes caused by pond drying and are more likely to accelerate development and forgo paedomorphosis for metamorphosis (Semlitsch and Wilbur 1989; Semlitsch et al. 1990). Although definitive differences in fitness as a function of differences in

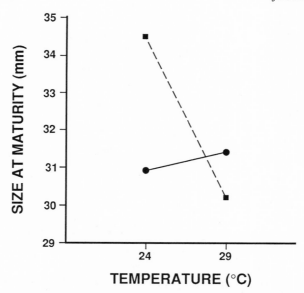

Fig. 9.5. Average standard length (mm) at maturity for female *P. formosa* as a function of temperature (°C) from clones derived from fish collected from two distinct drainages. Circles denote clones derived from the coastal area of Rio Tigre in northern Mexico, and squares denote clones derived from upland area of Rio Guayalejo. Each point represents averages of 10–14 females from data pooled across salinities of 12 and 20 parts per thousand at each temperature; there were no differences between length at maturity at these salinities at either temperature (see fig. 9.4, right). (Data from same source as figs. 9.1, 9.3, and 9.4)

plasticity have not been demonstrated, the evidence renders an adaptive explanation quite plausible.

There are many other examples of differences among conspecific populations in environmental sensitivity of life-history traits in both plants (Cook and Johnston 1968; Wilken 1977; Quinn and Hodgkinson 1984; Schlichting and Levin 1984; Scheiner and Goodnight 1984; Taylor and Aarsson 1988; Winn and Evans 1991) and animals (Bradshaw and Lounibos 1977; Stearns and Sage 1980; Bradshaw 1986; Conover and Heins 1987; Coyne and Beecham 1987; Etter 1988). Such distinctions are known for life-history traits that reflect ontogenetic development (fig. 9.5) as well as reproductive allocation (fig. 9.6). This roster reflects only cases that are candidates for adaptive differentiation and complements a comparably large roster of studies that demonstrate the existence of genetic variation within a population for plasticity that would provide the raw material that could be molded by differential selection in different conditions (Travis, in press).

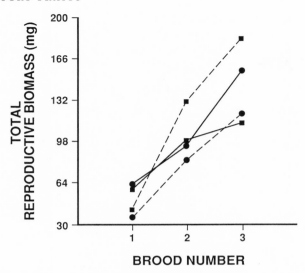

FIG. 9.6. Average reproductive biomass of female *P. formosa* as a function of female brood number (horizontal axis), geographic origin of clonal stock, and temperature. All data are derived from either 12 or 20 parts per thousand salinity (there were no differences between these values at these salinities). Each point is the average of 10–14 females per location; symbols for location as in figure 9.5; points connected by solid lines indicate the trajectory of reproductive biomass at a temperature of 29°C, and dashed lines indicate the trajectories at 24°C.

Several important caveats preclude a general conclusion that adaptive genetic divergence in plasticity is ubiquitous. First, although some studies have made precise genetic analyses of the response to a specific environmental effect like temperature (e.g., Coyne and Beecham 1987), others simply compare the response to variation in conditions that differ in many environmental features among samples derived from different populations without any formal genetic analysis. In these cases the factor(s) to which the organisms are responding is (are) unknown. Moreover, many such studies cannot distinguish a true genetic difference among the samples from induced maternal effects that may influence environmental sensitivity of progeny (Kinne 1962; Forster-Blouin 1989; Roach and Wulff 1987; Mousseau and Dingle 1991). Second, for only some of these examples can a testable hypothesis for the adaptive significance of the genetic differentiation for environmental sensitivity be offered (e.g., Bradshaw and Lounibos 1977). In others the ecological knowledge of the divergent populations is incomplete. Third, population differentiation for environmental sensitivity is not universal; several studies failed to uncover the phenomenon despite reasonable experimental designs (Scheiner et al. 1984; Brown 1985; Trexler et al. 1990). In at least one of these cases (Scheiner et al. 1984), the lack of differentiation in the pattern of phenotypic plasticity was contrary to a prediction for adaptive evolution.

The Proximate and Ultimate Genesis of Life-History Plasticity

General theories for the evolution of adaptive phenotypic plasticity in a variable environment have been reviewed elsewhere (Travis, in press; Via, chap. 4, this volume), as has been the state of knowledge about the genetic bases of plasticity (Jinks and Pooni 1988; Scheiner and Lyman 1991; Scheiner 1993). In this section I describe how phenotypic plasticity might best be integrated into general life-history theory.

The optimal life history is a problem of optimal allocation of energy among competing options that can include organismal maintenance, storage, growth, and reproduction (Sibly and Calow 1985; Dunham et al. 1989). Each option can contribute to fitness in a variety of ways. For example, energy stores may be drawn upon under stressful conditions to enhance the likelihood of survival (Newsome and Leduc 1975; Henderson et al. 1988) or they may be translated into increased reproductive output (Reznick and Braun 1987). Somatic growth can contribute to survivorship when it precedes the onset of conditions that produce size-dependent survival (Hunt 1969; Oliver et al. 1979; Conover and Ross 1982; Post and Evans 1989), but it can also contribute to reproductive output through a variety of size-fecundity relationships (Stearns and Koella 1986). The full solution to the problem of optimal allocation must be found by integrating each option's contribution to fitness, accounting for how that contribution is actually made.

Several models based on optimal control theory have been developed to do this (e.g., Taylor et al. 1974; Iwasa and Roughgarden 1984; Sibly and Calow 1986), and Perrin's (1992) paper generalizes these approaches. The optimal life history allocates energy among prereproductive options in proportion to their expected return in fitness as long as the return for each option exceeds the investment. The energy is allocated first to the option that gives the highest marginal value for the investment; when the return from the first option decreases to the expected return from a second option, then investment is expanded to the second option such that the two options provide equal returns. Nonuniform allocations may readily occur because it is the marginal value of the options for fitness that is equalized and not the investment level per se; indeed nonuniform allocations can form the theoretical basis for allometric growth (Perrin 1992). Reproduction should begin when the total rate of return in somatic structures decreases to the level of investment; in simpler models all subsequently assimilated energy is allocated to reproduction (Taylor et al. 1974). In fuller models the optimal balance between investment in reproduction and further investment in somatic options is a function of shape of the relationship between somatic growth and reproduction or mortality (Sibly et al. 1985) or the variability in the demographic environment (Cohen 1971; King and Roughgarden 1982).

These results offer a framework for linking phenotypic plasticity and life-history theory. First, in a variable environment the linkage between one or more allocation options and fitness will vary. For example, the linkage between larval period and survival in anurans changes with rainfall pattern (Newman 1988). Second, the relationship between somatic growth and maintenance and reproductive investment will vary with the amount of resources available because a strict, forceful trade-off may only occur under energetically limiting conditions (Dunham et al. 1989). Third, for many populations the demographic environment is highly variable from generation to generation (Ferguson and Fox 1984; Schmidt and Levin 1985; Winn and Werner 1987) and patterns of reproduction are often altered concordantly (Nichols et al. 1976). These points imply that in an environment that varies temporally with respect to sources and rates of mortality and in energetic availability, the optimal allocation to somatic and reproductive functions will also vary.

When the optimal life history does vary, phenotypic plasticity or protected genetic polymorphisms may be the evolutionary product. General models of plasticity have not offered clear, explicit guidelines on when each outcome might be expected. Obviously adaptive plasticity can occur only if a reasonably reliable cue is available before allocations are made that will indicate the particular environmental condition. If such a cue exists, and if there is genetic variation that produces variable trait expression (here the trait is, in theory, an allocation pattern), then an optimal pattern of plasticity can evolve. If no cue exists, enhanced phenotypic variation might evolve in some cases (McGinley et al. 1987; Bull 1987 and references therein). Although there is no full, formal theory along these lines for the evolution of adaptive life-history plasticity per se, elements of such a theory have been offered by several workers (e.g., Stearns and Koella 1986; Venable and Brown 1988).

Qualitatively, which criteria must be fulfilled before any particular example of plasticity in life-history traits might be diagnosed as adaptive? First, the optimum value of the trait(s) in question must vary among environments, and the direction of the change produced by plasticity among environments must be at least in the direction of the change in the optimum trait value. For example, if the optimum age at maturity in an energetically limiting environment is later than the optimum for an environment that is not energetically limiting, then the plasticity in age at maturity must be in the direction of delaying maturation.

This criterion is not so easily examined in any empirical situation. Selection will act on the entire pattern of allocation indirectly through the connections of all of the affected life-history traits with fitness, as described in the optimal control theory models. This means that the adaptive significance of life-history plasticity will require an evaluation of the optimum multivariate phenotype in each condition. For example, it may never be advantageous to delay maturity, but in an energetically stressful environment it may be advantageous to reproduce at a large size. If the energetic limitations of the environment produce low growth rates and if age and size at maturity are genetically connected via inescapable pleiotropy, then the best phenotype may be the one that "endures" the delayed maturity in exchange for the benefits of large size. The plasticity in age at maturity has no adaptive significance and may even produce a less fit phenotypic value when this trait is considered alone. Thus the full spectrum of direct and indirect selection pressures may need to be evaluated in more complex cases before a diagnosis of adaptive plasticity can be made.

This first criterion embraces the theoretical criterion for distinct optimum phenotypes in distinct environments, the multivariate nature of fitness when individual life-history traits are considered, and the presence of an informative cue so that the development of the phenotype can be modified appropriately. For life-history traits a second criterion is necessary: the plasticity in individual traits must reflect a fundamental change in the allocation pattern. This criterion arises because a life history is, by definition, a pattern of energy allocation among competing options for somatic and reproductive use; that pattern is molded by indirect selection through the direct selection on trait values and the physiological compromises that are made among allocation options to produce those trait values. If the pattern of energy allocation is unchanged across environments, then the life history has not changed across environments; in this case the plasticity of individual traits represents merely a proportionate change in the total energy flux through the population (Dunham et al. 1989).

This second criterion is analogous to Smith-Gill's (1983) assertion that adaptive phenotypic plasticity can be diagnosed only when there is clear evidence for a flexible developmental pattern and not simply evidence for trait variation in response to environmental variation (Reznick 1990). The criterion might eliminate many of the examples cited in the previous two sections from consideration as adaptive. For example, the increased brood size of lecithotrophic poeciliids in response to food level increases (e.g., Trendall 1983) may reflect only a change in energy flux through the population and not a change in the allocation of energy (cf. Reznick

1983). However, the alteration in the correlation between offspring number and size that Dahlgren (1980) reported from his studies of density variation appears to represent a clear change in allocation patterns, as does the alteration between lecithotrophic and matrotrophic provisioning of embryos that was reported by Trexler (1985). Alterations in growth and development patterns in juvenile guppies (Reznick 1990) and larval anurans (Wilbur and Collins 1973) in response to environmental variation also appear to represent changes in the allocation of energy between somatic growth and the demands of developmental differentiation.

Conclusions

Evolutionary biology has made enormous strides toward the goal of understanding which traits are most likely to evolve in different ecological settings. Theoretical and empirical studies have indicated that there is a fundamental, causal linkage between the patterns of mortality rates and the optimum schedule of reproductive timing, investment, and packaging. Many empirical examples of life-history variation can be accounted for by this linkage, whether they involve variation at the species level or genetically based variation among conspecific populations. However, many examples cannot be easily reconciled in this fashion; in these cases there is always an element of environmental influence on the phenotypic variation that obscures the putative link with the known demographic variation. The incorporation of this phenomenon in nontrivial fashion into an understanding of life-history diversity represents the emerging challenge in the discipline.

Conspecific populations often exhibit distinct patterns of phenotypic plasticity in life-history traits; these cases offer the most suitable testing ground for elucidating whether such plasticity is itself a product of natural selection. An adaptive pattern of plasticity can be diagnosed only if two major criteria are met. One criterion involves the correspondence between realized trait values and the actual optimum trait values for each condition experienced by the population. The other criterion is that phenotypic plasticity in individual life-history traits must reflect real plasticity in the allocation of energy to competing options and not just a proportional change in energy flux through a population.

Both criteria may be difficult to verify in specific cases because both involve the multivariate nature of the life-history phenotype. With respect to the first criterion, the contribution of different individual life-history traits to fitness will certainly change with the changes in ecological condition. For example, although the age at first reproduction is a crucial determinant of fitness in a continuously growing population, it makes little contribution when individuals live for very long periods and make low annual investments; in this situation the average annual fecundity plays a larger role (Caswell 1982).

For the second criterion, several traits will have to be examined as a function of environmental variation in order to ascertain how (if at all) the allocation of energy is altered. In many cases this problem might best be addressed with a more physiological approach than is usually taken by ecological geneticists, because this approach might better reveal the nature of allocation decisions than the indirect measures of individual life-history traits. Such a blend of physiology and genetics would complement analogous laboratory studies of the physiological basis of differential allocation patterns in *Drosophila* that produce different life histories (Partridge and Andrews 1985; Service 1987; Luckinbill et al. 1988) and bring evolutionary biologists closer to a general understanding of life-history variation in nature.

Acknowledgments

I thank Les Real for this opportunity and especially for his infinite patience with a shockingly slow author. I owe an enormous debt to my colleagues D. Reznick and N. Perrin; in many ways they have educated me and guided my thinking along avenues I would not have otherwise explored. I am also indebted to my colleagues and students J. C. Trexler and M. McManus; they have also taught me a great deal and their ideas and work have inspired my own. I thank A. Winn for her insightful comments on a previous draft of this manuscript. I was supported during the writing of this paper by NSF grant BSR 88-18001.

References

Antonovics, J., and R. B. Primack. 1982. Experimental ecological genetics in *Plantago*. VI. The demography of seedling transplants of *P. lanceolata. J. Ecol.* 70: 55–75.
Avise, J. C., J. C. Trexler, J. Travis, and W. S. Nelson. 1991. *Poecilia mexicana* is the recent female parent of the unisexual fish *P. formosa. Evolution* 45: 1530–1533.
Bagenal, T. B. 1969. The relationship between food supply and fecundity in brown trout *Salmo trutta* L. *J. Fish Biol.* 1: 176–182.
Bagenal, T. B. 1971. The interrelation of the size of eggs, the date of spawning and the production cycle. *J. Fish Biol.* 3: 207–219.
Baker, H. G. 1972. Seed weight variation in relation to environmental conditions in California. *Ecology* 53: 997–1010.

Ballinger, R. E. 1973. Comparative demography of two viviparous iguanid lizards (*Sceloporus jarrovi* and *Sceloporus poinsettii*). *Ecology* 54: 269–283.

Bell, G. 1989. A comparative method. *Amer. Nat.* 133: 553–571.

Bergmans, M. 1984. Life history adaptation to demographic regime in laboratory-cultured *Tisbe furcata* (Copepoda, Harpacticoida). *Evolution* 38: 292–299.

Berven, K. A. 1982a. The genetic basis of altitudinal variation in the wood frog, *Rana sylvatica*. I. An experimental analysis of life history traits. *Evolution* 36: 962–983.

Berven, K. A. 1982b. The genetic basis of altitudinal variation in the wood frog, *Rana sylvatica*. II. An experimental analysis of larval traits. *Oecologia* 52: 360–369.

Berven, K. A. 1987. The heritable basis of variation in larval developmental patterns within populations of the wood frog (*Rana sylvatica*). *Evolution* 41: 1088–1097.

Berven, K. A., D. E. Gill, and S. J. Smith-Gill. 1979. Countergradient selection in the green frog, *Rana clamitans*. *Evolution* 33: 609–623.

Boyce, M. S. 1984. Restitution of r- and K-selection as a model of density-dependent natural selection. *Ann. Rev. Ecol. Syst.* 15: 427–447.

Bradshaw, A. D. 1965. Evolutionary significance of phenotypic plasticity in plants. *Adv. in Genetics* 13: 115–155.

Bradshaw, W. E. 1986. Variable iteroparity as a life-history tactic in the pitcher-plant mosquito *Wyeomyia smithii*. *Evolution* 40: 471–478.

Bradshaw, W. E., and L. P. Lounibos. 1977. Evolution of dormancy and its photoperiodic control in pitcher-plant mosquitos. *Evolution* 31: 546–567.

Brewer, R., and L. Swander. 1977. Life history factors affecting the intrinsic rate of natural increase of birds of the deciduous forest biome. *Wilson Bulletin* 89: 211–232.

Brooks, D. R., and D. A. McLennan. 1991. *Phylogeny, Ecology, and Behavior*. University of Chicago Press, Chicago.

Brown, K. M. 1979. The adaptive demography of four freshwater pulmonate snails. *Evolution* 33: 417–432.

Brown, K. M. 1983. Do life history tactics exist at the intraspecific level? Data from freshwater snails. *Amer. Nat.* 121: 871–879.

Brown, K. M. 1985. Intraspecific life history variation in a pond snail: The roles of population divergence and phenotypic plasticity. *Evolution* 39: 387–395.

Browne, R. A., and W. D. Russell-Hunter. 1978. Reproductive effort in molluscs. *Oecologia* 37: 23–27.

Bull, J. J. 1987. Evolution of phenotypic variance. *Evolution* 41: 303–315.

Burt, A. 1989. Comparative methods using phylogenetically independent contrasts. *Oxford Surveys in Evol. Biol.* 6: 33–53.

Busack, C. A., and G.A.E. Gall. 1983. An initial description of the quantitative genetics of growth and reproduction in the mosquitofish, *Gambusia affinis*. *Aquaculture* 32: 123–140.

Caswell, H. 1982. Life history and the equilibrium status of populations. *Amer. Nat.* 120: 317–339.

Charlesworth, B. 1980. *Evolution in Age-structured Populations*. Cambridge University Press, Cambridge, U.K.

Charnov, E. L. 1991. Evolution of life history variation among female mammals. *Proc. Natl. Acad. Sci. USA* 88: 1134–1137.

Cheverud, J. M., M. M. Dow, and W. Leutenegger. 1985. The quantitative assessment of phylogenetic constraints in comparative analyses: Sexual dimorphism in body weight among primates. *Evolution* 39: 1335–1351.

Clark, A. G. 1987. Senescence and the genetic correlation hang-up. *Amer. Nat.* 129: 932–940.

Cody, M. L. 1966. A general theory of clutch size. *Evolution* 20: 174–184.

Cody, M. L. 1971. Ecological aspects of reproduction. In *Avian Biology*, ed. D. S. Farner and J. R. King, pp. 462–512. Academic Press, New York.

Cohen, D. 1971. Maximizing final yield when growth is limited by time or by limiting resources. *J. Theoret. Biol.* 33: 299–307.

Conover, D. O., and S. W. Heins. 1987. Adaptation variation in environmental and genetic sex determination in a fish. *Nature* 326: 496–498.

Conover, D. O., and T.M.C. Present. 1990. Countergradient variation in growth rate: Compensation for length of the growing season among Atlantic silversides from different latitudes. *Oecologia* 83: 316–324.

Conover, D. O., and M. R. Ross. 1982. Patterns in seasonal abundance, growth, and biomass of the Atlantic silverside, *Menidia menidia*, in a New England estuary. *Estuaries* 5: 275–286.

Constantz, G. D. 1979. Life history patterns of a livebearing fish in contrasting environments. *Oecologia* 40: 189–201.

Cook, R. E. 1975. The photoinductive control of seed weight in *Chenopodium rubrum* L. *Amer. J. Bot.* 62: 427–431.

Cook, S. A., and M. P. Johnston. 1968. Adaptation to heterogeneous environments. I. Variation in heterophylly in *Ranunculus flammula* L. *Evolution* 22: 496–516.

Coyne, J. A., and E. Beecham. 1987. Heritability of two morphological characters within and among natural populations of *Drosophila melanogaster*. *Genetics* 117: 727–737.

Crowl, T. A., and A. P. Covich. 1990. Predator-induced life-history shifts in a freshwater snail. *Science* 247: 949–951.

Dahlgren, B. T. 1979. The effects of population density on fecundity and fertility in the guppy, *Poecilia reticulata* (Peters). *J. Fish Biol.* 15: 71–91.

Dahlgren, B. T. 1980. Influences of population density on reproductive output at food excess in the guppy, *Poecilia reticulata* (Peters). *Biol. Reproduction* 22: 1047–1061.

Dingle, H. 1981. Geographic variation and behavioral flexibility in milkweed bug life histories. In *Insects and Life History Patterns: Geographic and Habitat Variation*, ed. R. F. Denno and H. Dingle, pp. 57–73. Springer-Verlag, New York.

Dingle, H., and J. D. Baldwin. 1983. Geographic variation in life histories: A comparison of tropical and temperate milkweed bugs. In *Diapause and Life Cycle Strategies in Insects*, ed. V. K. Brown and I. Hodek, pp. 143–166. Junk, The Hague.

Dingle, H., B. M. Alden, N. R. Blakley, D. Kopec, and E. R. Miller. 1980. Variation in photoperiodic response within and among species of milkweed bugs (*Oncopeltus*). *Evolution* 34: 356–370.

Dingle, H., W. S. Blau, C. K. Brown, and J. P. Hegmann. 1982. Population crosses and the genetic structure of milkweed bug life histories. In *Evolution and Genetics of Life Histories*, ed. H. Dingle and J. P. Hegmann, pp. 209–227. Springer-Verlag, New York.

Dunham, A. E. 1982. Demographic and life history variation among populations of the iguanid lizard *Urosaurus ornatus*: Implications for the study of life-history phenomena in lizards. *Herpetologica* 38: 208–221.

Dunham, A. E., D. B. Miles, and D. N. Reznick. 1988. Life history patterns in squamate reptiles. *Biology of the Reptilia*, ed. C. Gans and R. B. Huey, pp. 441–522. Vol. 16, Ecology B, *Defense and Life History*. Liss, New York.

Dunham, A. E., B. W. Grant, and K. L. Overall. 1989. Interfaces between biophysical and physiological ecology and the population ecology of terrestrial vertebrate ectotherms. *Physiol. Zool.* 62: 335–355.

Elgar, M. A. 1990. Evolutionary compromise between a few large and many small eggs: Comparative evidence in teleost fish. *Oikos* 59: 283–287.

Elgar, M. A., and L. J. Heaphy. 1989. Covariation between clutch size, egg weight and egg shape: Comparative evidence for chelonians. *J. Zool. (Lond.)* 219: 137–152.

Etter, R. J. 1988. Asymmetrical developmental plasticity in an intertidal snail. *Evolution* 42: 322–334.

Etter, R. J. 1989. Life history variation in the intertidal snail *Nucella lapillus* across a wave-exposure gradient. *Ecology* 70: 1857–1876.

Felsenstein, J. 1988. Phylogenies and quantitative characters. *Ann. Rev. Ecol. Syst.* 19: 445–471.

Ferguson, G. W., and S. F. Fox. 1984. Annual variation of survival advantage of large juvenile side-blotched lizards, *Uta stansburiana*: Its causes and evolutionary significance. *Evolution* 38: 342–349.

Ford, N. B., and R. A. Seigel. 1989. Phenotypic plasticity in reproductive traits: Evidence from a viviparous snake. *Ecology* 70: 1768–1774.

Forster-Blouin, S. 1989. Genetic and Environmental Components of Thermal Tolerance in the Least Killifish, *Heterandria formosa*. Ph.D. dissertation, Florida State University, Tallahassee.

Fowler, N. L., and J. Antonovics. 1981. Small-scale variability in the demography of transplants of two herbaceous species. *Ecology* 62: 1450–1457.

Gadgil, M., and O. T. Solbrig. 1972. The concept of r- and K-selection: Evidence from wild flowers and some theoretical considerations. *Amer. Nat.* 106: 14–31.

Gaines, M. S., K. J. Vogt, J. L. Hamrick, and J. Caldwell. 1974. Reproductive strategies and growth patterns in sunflowers (*Helianthus*). *Amer. Nat.* 108: 889–894.

Giesel, J. T. 1986. Genetic correlation structure of life history variables in outbred, wild *Drosophila melanogaster*: Effects of photoperiod regimen. *Amer. Nat.* 128: 593–603.

Giesel, J. T., P. A. Murphy, and M. N. Manlove. 1982. The influence of temperature on genetic interrelationships of life history traits in a population of *Droso-*

phila melanogaster: What tangled data sets we weave. *Amer. Nat.* 119: 464–479.

Goodman, D. 1982. Optimal life histories, optimal notation, and the value of reproductive value. *Amer. Nat.* 119: 803–823.

Grafen, A. 1989. The phylogenetic regression. *Phil. Trans. Roy. Soc. London B* 326: 119–157.

Groeters, F. R., and H. Dingle. 1987. Genetic and maternal influences on life history plasticity in response to photoperiod by milkweed bugs (*Oncopeltus fasciatus*). *Amer. Nat.* 129: 332–346.

Guyer, C. 1988. Food supplementation in a tropical mainland anole, *Norops humilis*: Effects on individuals. *Ecology* 69: 362–369.

Harris, R. N. 1987. Density-dependent paedomorphosis in the salamander *Notophthalmus viridescens dorsalis*. *Ecology* 68: 705–712.

Healey, M. C., and K. Dietz. 1984. Variation in fecundity of lake whitefish (*Coregonus clupeaformis*) from Lesser Slave and Utijuma lakes in northern Alberta. *Copeia* 1984: 238–242.

Henderson, P. A., R.H.A. Holmes, and R. N. Bamber. 1988. Size-selective overwintering mortality in the sand smelt, *Atherina boyeri* Risso, and its role in population regulation. *J. Fish Biol.* 33: 221–233.

Hislop, J.R.G., A. P. Robb, and J. A. Gauld. 1978. Observations on effects of feeding level on growth and reproduction in haddock, *Melanogrammus aeglefinus* (L.), in captivity. *J. Fish Biol.* 13: 85–98.

Hubbs, C., M. M. Stevenson, and A. E. Peden. 1968. Fecundity and egg size in two central Texas darter populations. *Southwest Nat.* 13: 301–324.

Hunt, R. L. 1969. Overwinter survival of wild fingerling brook trout in Lawrence Creek, Wisconsin. *J. Fish. Res. Board Canada* 35: 1473–1483.

Iwasa, Y., and J. Roughgarden. 1984. Shoot/root balance of plants: Optimal growth of a system with many vegetative organs. *Theoret. Pop. Biol.* 25: 78–105.

Iwasa, Y., and E. Teramoto. 1980. A criterion of life history evolution based on density-dependent selection. *J. Theoret. Biol.* 13: 1–68.

James, F. C. 1991. Complementary descriptive and experimental studies of clinal variation in birds. *Amer. Zool.* 31: 694-706.

Jinks, J. L., and H. S. Pooni. 1988. The genetic basis of environmental sensitivity. In *Proceedings of the Second International Conference on Quantitative Genetics*, ed. B. S. Weir, E. J. Eisen, M. M. Goodman, and G. Namkoong, pp. 505–522. Sinauer, Sunderland, Mass.

Kaplan, R. H. 1987. Developmental plasticity and maternal effects of reproductive characteristics in the frog, *Bombina orientalis*. *Oecologia* 71: 273–279.

Kaplan, R. H., and S. N. Salthe. 1979. The allometry of reproduction: An empirical view in salamanders. *Amer. Nat.* 113: 671–689.

Keen, W. H., J. Travis, and J. Juilianna. 1984. Larval growth in three sympatric *Ambystoma* salamander species: Species differences and effects of temperature. *Canadian J. Zool.* 62: 1043–1047.

King, D., and J. Roughgarden. 1982. Graded allocation between vegetative and reproductive growth for annual plants in growing seasons of random length. *Theoret. Pop. Biol.* 22: 1–16.

Kinne, O. 1962. Irreversible nongenetic adaptation. *Comp. Biochem. Physiol.* 5: 265–282.

Lacey, E. P. 1986. The genetic and environmental control of reproductive timing in a short-lived monocarpic species *Daucus carota* (Umbelliferae). *J. Ecol.* 74: 73–86.

Lande, R. 1982. A quantitative genetic theory of life history evolution. *Ecology* 63: 607–615.

Law, R. 1979. The cost of reproduction in annual meadow grass. *Amer. Nat.* 113: 3–16.

Law, R., A. D. Bradshaw, and P. D. Putwain. 1977. Life-history variation in *Poa annua*. *Evolution* 31: 233–246.

Leggett, W. C., and J. E. Carscadden. 1978. Latitudinal variation in reproductive characteristics of American shad (*Alosa sapidissima*): Evidence for population specific life history strategies in fish. *J. Fish. Res. Board Canada* 35: 1469–1478.

Luckinbill, L. S. 1978. r- and K-selection in experimental populations of *Escherichia coli*. *Science* 202: 1201–1203.

Luckinbill, L. S. 1984. An experimental analysis of a life history theory. *Ecology* 65: 1170–1184.

Luckinbill, L. S., and M. J. Clare. 1986. A density threshold for the expression of longevity in *Drosophila melanogaster*. *Heredity* 56: 529–535.

Luckinbill, L. S., R. Arking, M. G. Clare, W. C. Cirocco, and S. A. Buck. 1984. Selection for delayed senescence in *Drosophila melanogaster*. *Evolution* 38: 996–1003.

Luckinbill, L. S., J. L. Graves, A. Tomkiw, and O. Srowirka. 1988. A qualitative analysis of some life-history correlates of longevity in *Drosophila melanogaster*. *Evol. Ecol.* 2: 85–94.

MacArthur, R. H., and E. O. Wilson. 1967. *The Theory of Island Biogeography*. Princeton University Press, Princeton, N.J.

McGinley, M. A., D. H. Temme, and M. A. Geber. 1987. Parental investment in offspring in variable environments: Theoretical and empirical considerations. *Amer. Nat.* 130: 370–398.

McWilliams, E. L., R. Q. Landers, and J. P. Mahlstede. 1968. Variation in seed weight and germination in populations of *Amaranthus retroflexus* L. *Ecology* 49: 290–296.

May, R. M., and D. Rubenstein. 1985. Reproductive strategies. In *Reproductive Fitness*, ed. R. Short and M. V. Austin, vol. 4, pp. 1–23. Cambridge University Press, Cambridge, U.K.

Mazer, S. J. 1989. Ecological, taxonomic, and life history correlates of seed mass among Indiana dune angiosperms. *Ecol. Monogr.* 59: 153–175.

Mazer, S. J., and C. T. Schick. 1991. Constancy of population parameters for life-history and floral traits in *Raphanus sativus* L. II. Effects of planting density on phenotype and heritability estimates. *Evolution* 45: 1888–1907.

Michod, R. E. 1979. Evolution of life histories in response to age-specific mortality factors. *Amer. Nat.* 1113: 531–550.

Miles, D. B., and A. E. Dunham. 1992. Comparative analyses of phylogenetic effects in the life history patterns of iguanid reptiles. *Amer. Nat.* 139: 848–869.

Millar, J. S., and R. M. Zammuto. 1983. Life histories of mammals: An analysis of life tables. *Ecology* 64: 631–635.

Mills, C. A. 1988. The effect of extreme northerly climatic conditions on the life history of the minnow, *Phoxinus phoxinus* (L.). *J. Fish Biol.* 33: 545–561.

Mitton, J. B., and W. M. Lewis, Jr. 1989. Relationships between genetic variability and life-history features of bony fishes. *Evolution* 43: 1712–1723.

Mousseau, T. A., and H. Dingle. 1991. Maternal effects in insect life histories. *Ann. Rev. Entomol.* 36: 511–534.

Mueller, L. D. 1988. Density-dependent population growth and natural selection in food-limited environments: The *Drosophila* model. *Amer. Nat.* 132: 786–809.

Murphy, P. A., J. T. Giesel, and M. N. Manlove. 1983. Temperature effects on life history variation in *Drosophila simulans*. *Evolution* 37: 1181–1192.

Newman, R. A. 1988. Adaptive plasticity in development of *Scaphiopus couchii* tadpoles in desert ponds. *Evolution* 42: 774–783.

Newsome, G. E., and G. Leduc. 1975. Seasonal changes of fat content in the yellow perch (*Perca flavescens*) of two Lawrentian lakes. *J. Fish. Res. Board Canada* 32: 2214–2221.

Nichols, J. D., W. Conley, B. Batt, and A. R. Tipton. 1976. Temporally dynamic reproductive strategies and the concept of r- and K-selection. *Amer. Nat.* 110: 995–1005.

Nussbaum, R. A. 1981. Seasonal shifts in clutch-size and egg-size in the side-blotched lizard, *Uta stansburiana*, Baird and Girard. *Oecologia* 49: 8–13.

Oliver, J. D., G. F. Holeton, and K. F. Chua. 1979. Overwinter mortality of fingerling smallmouth bass in relation to size, relative energy stores, and environmental temperature. *Trans. Amer. Fish. Soc.* 108: 130–136.

Pagel, M., and P. Harvey. 1988. Recent developments in the analysis of comparative data. *Quart. Rev. Biol.* 63: 413–440.

Palmer, J. O., and H. Dingle. 1986. Direct and correlated responses to selection among life-history traits in milkweed bugs (*Oncopeltus fasciatus*). *Evolution* 40: 767–777.

Parenti, L. R., and M. Rauchenberger. 1989. Systematic overview of the poeciliines. In *Ecology and Evolution of Livebearing Fishes (Poeciliidae)*, ed. G. K. Meffe and F. F. Snelson, Jr., pp. 3–12. Prentice Hall, New York.

Partridge, L., and R. Andrews. 1985. The effect of reproductive activity on the longevity of male *Drosophila melanogaster* is not caused by an acceleration of ageing. *J. Insect Physiol.* 31: 393–395.

Perrin, N. 1988. Why are offspring born larger when it is colder? Phenotypic plasticity for offspring size in the cladoceran *Simocephalus vetulus* (Muller). *Functional Ecol.* 2: 283–288.

Perrin, N. 1992. Optimal resource allocation and the marginal value of organs. *Amer. Nat.* 139: 1344–1369.

Pianka, E. R. 1970. On r- and K-selection. *Amer. Nat.* 104: 592–596.

Platenkamp, G.A.J. 1990. Phenotypic plasticity and genetic differentiation in the demography of the grass *Anthoxanthum odoratum*. *J. Ecol.* 78: 772–788.

Platenkamp, G.A.J. 1991. Phenotypic plasticity and population differentiation in seeds and seedlings of the grass *Anthoxanthum odoratum. Oecologia* 88: 515–520.

Post, J. R., and D. O. Evans. 1989. Size-dependent overwinter mortality of young-of-the-year yellow perch (*Perca flavescens*): Laboratory, in situ enclosure and field experiments. *Canadian J. Fish. Aquat. Sci.* 46: 1958–1960.

Primack, R. B. 1979. Reproductive effort in annual and perennial species of *Plantago* (Plantaginaceae). *Amer. Nat.* 114:51–62.

Primack, R. B. 1985. Longevity of individual flowers. *Ann. Rev. Ecol. Syst.* 16: 15–37.

Primack, R. B. 1987. Relationships among flowers, fruits, and seeds. *Ann. Rev. Ecol. Syst.* 18: 409–430.

Primack, R. B., and J. Antonovics. 1982. Experimental ecological genetics in *Plantago*. VII. Reproductive effort in populations of *P. lanceolata* L. *Evolution* 36: 742–752.

Pugliese, A., and J. Kozlowski. 1990. Optimal patterns of growth and reproduction for perennial plants with persisting and or not persisting vegetative parts. *Evol. Ecol.* 4: 75–89.

Quinn, J. A., and K. C. Hodgkinson. 1984. Population variability in *Danthonia caespitosa* (Gramineae) in response to increasing density under three temperature regimes. *Amer. J. Bot.* 70: 1425–1431.

Read, A. F., and P. H. Harvey. 1989. Life history differences among the eutherian radiations. *J. Zool. (London)* 219: 329–353.

Reinartz, J. A. 1984a. Life history of common mullein (*Verbascum thapsus*). I. Latitudinal differences in population dynamics and timing of reproduction. *J. Ecol.* 72: 897–912.

Reinartz, J. A. 1984b. Life history of common mullein (*Verbascum thapsus*). II. Plant size, biomass partitioning and morphology. *J. Ecol.* 72: 913–925.

Reinartz, J. A. 1984c. Life history of common mullein (*Verbascum thapsus*). III. Differences among sequential cohorts. *J. Ecol.* 72: 927–936.

Reznick, D. 1982a. The impact of predation on life history evolution in Trinidadian guppies: Genetic basis of observed life history patterns. *Evolution* 36: 1236–1250.

Reznick, D. 1982b. Genetic determination of offspring size in the guppy (*Poecilia reticulata*). *Amer. Nat.* 120: 181–188.

Reznick, D. 1983. The structure of guppy life histories: The tradeoff between growth and reproduction. *Ecology* 64: 862–873.

Reznick, D. N. 1989. Life history evolution in guppies. 2. Repeatability of field observations and the effects of season on life histories. *Evolution* 43: 1285–1297.

Reznick, D. N. 1990. Plasticity in age and size at maturity in male guppies (*Poecilia reticulata*): An experimental evaluation of alternative models of development. *J. Evol. Biol.* 3: 185–203.

Reznick, D. N., and B. Braun. 1987. Fat cycling in the mosquitofish (*Gambusia affinis*): Is fat storage a reproductive adaptation? *Oecologia* 73: 401–413.

Reznick, D. N., and H. Bryga. 1987. Life history evolution in guppies (*Poecilia reticulata*). 1. Phenotypic and genetic changes in an introduction experiment. *Evolution* 41: 1370–1385.

Reznick, D., and J. A. Endler. 1982. The impact of predation on life history evolution in Trinidadian guppies (*Poecilia reticulata*). *Evolution* 36: 160–177.

Reznick, D. N., and D. B. Miles. 1989. A review of life history patterns in poeciliid fishes. In *Ecology and Evolution of Livebearing Fishes (Poeciliidae)*, ed. G. K. Meffe and F. F. Snelson, Jr., pp. 125–148. Prentice Hall, New York.

Reznick, D. A., H. Bryga, and J. A. Endler. 1990. Experimentally induced life-history evolution in a natural population. *Nature* 346: 357–359.

Roach, D. A., and R. D. Wulff. 1987. Maternal effects in plants. *Ann. Rev. Ecol. Syst.* 18: 209–236.

Rohwer, F. C. 1988. Inter- and intraspecific relationships between egg size and clutch size in waterfowl. *Auk* 105: 161–176.

Rose, M. R. 1982. Antagonistic pleiotropy, dominance, and genetic variation. *Heredity* 48: 63–78.

Rose, M. R. 1985. Life history evolution with antagonistic pleiotropy and overlapping generations. *Theoret. Pop. Biol.* 28: 342–358.

Rose, M. R., and B. Charlesworth. 1981. Genetics and life history in *Drosophila melanogaster*. II. Exploratory selection experiments. *Genetics* 97: 187–196.

Rubenstein, D. I. 1981. Individual variation and competition in the Everglades pygmy sunfish. *J. Anim. Ecol.* 50: 337–350.

Saether, B.-E. 1988. Pattern of covariation between life-history traits of European birds. *Nature* 331: 616–617.

Schaffer, W. M. 1983. The application of optimal control theory to the general life history problem. *Amer. Nat.* 121: 418–431.

Schaffer, W. M., and P. F. Elson. 1975. The adaptive significance of variations in life history among local populations of Atlantic salmon in North America. *Ecology* 56: 577–590.

Scheiner, S. 1993. In press. The genetics and evolution of phenotypic plasticity. *Ann. Rev. Ecol. Syst.* 24.

Scheiner, S., and C. J. Goodnight. 1984. The comparison of phenotypic plasticity and genetic variation in populations of the grass *Danthonia spicata*. *Evolution* 38: 845–855.

Scheiner, S., and R. F. Lyman. 1991. The genetics of phenotypic plasticity. II. Response to selection. *J. Evol. Biol.* 4: 23–50.

Scheiner, S. M., J. Gurevitch, and J. Teeri. 1984. A genetic analysis of the photosynthetic properties of populations of *Danthonia spicata* that have different growth responses to light level. *Oecologia* 64: 74–77.

Schemske, D. W. 1984. Population structure and local selection in *Impatiens pallida* (Balsaminaceae), a selfing annual. *Evolution* 38: 817–832.

Schimel, D. S. 1992. Population and community processes in the response of global ecosystems to global change. In *Biotic Interactions and Global Change*, ed. R. B. Huey, P. Kareiva, J. Kingsolver, T. Root, and F. S. Chapin, pp. 45–54. Sinauer, Sunderland, Mass.

Schimpf, D. J. 1977. Seed weight of *Amaranthus retroflexus* in relation to moisture and length of growing season. *Ecology* 58: 450–453.

Schlichting, C. D. 1986. The evolution of phenotypic plasticity in plants. *Ann. Rev. Evol. Syst.* 17: 667–693.

Schlichting, C. D., and D. A. Levin. 1984. Phenotypic plasticity of annual phlox: Tests of some hypotheses. *Amer. J. Bot.* 7: 252–260.

Schmidt, K. P., and D. A. Levin. 1985. The comparative demography of reciprocally sown populations of *Phlox drummondii* Hook. I. Survivorships, fecundities, and finite rates of increase. *Evolution* 39: 396–404.

Schoenherr, A. A. 1977. Density dependent and density independent regulation of reproduction in the Gila topminnow, *Poeciliopsis occidentalis* (Baird and Girard). *Ecology* 58: 438–444.

Semlitsch, R. D. 1987a. Paedomorphosis in *Ambystoma talpoideum*: Effects of density, food, and pond drying. *Ecology* 68: 994–1002.

Semlitsch, R. D. 1987b. Density-dependent growth and fecundity in the paedomorphic salamander *Ambystoma talpoideum*. *Ecology* 63: 1003–1008.

Semlitsch, R. D., and J. W. Gibbons. 1985. Phenotypic variation in metamorphosis and paedomorphosis in the salamander *Ambystoma talpoideum*. *Ecology* 66: 1123–1130.

Semlitsch, R. D., and H. M. Wilbur. 1989. Artificial selection for paedomorphosis in the salamander *Ambystoma talpoideum*. *Evolution* 43: 105–112.

Semlitsch, R. D., R. N. Harris, and H. M. Wilbur. 1990. Paedomorphosis in *Ambystoma talpoideum*: Maintenance of population variation and alternative life-history pathways. *Evolution* 44: 1604–1613.

Service, P. M. 1987. Physiological mechanisms of increased stress resistance in *Drosophila melanogaster* selected for postponed senescence. *Physiol. Zool.* 60: 321–326.

Sibly, R. M., and P. Calow. 1985. *Physiological Ecology of Animals: An Evolutionary Approach.* Blackwell, Oxford.

Sibly, R. M., and P. Calow. 1986. Growth and resource allocation. In *Evolutionary Physiological Ecology*, ed. P. Calow, pp. 37–52. Cambridge University Press, Cambridge, U.K.

Sibly, R., P. Calow, and N. Nichols. 1985. Are patterns of growth adaptive? *J. Theoret. Biol.* 112: 553–574.

Skelly, D. K., and E. E. Werner. 1990. Behavioral and life-historical responses of larval American toads to an odonate predator. *Ecology* 71: 2313–2322.

Smith, B. H. 1983. Demography of *Floerkea proserpinacoides*, a forest-floor annual. II. Density-dependent reproduction. *J. Ecol.* 71: 405–412.

Smith-Gill, S. J. 1983. Developmental plasticity: Developmental conversion versus phenotypic modulation. *Amer. Zool.* 23: 47–56.

So, P.-M., and A. Takafuji. 1991. Coexistence of *Tetranychus urticae* (Acarina: Tetranychidae) with different capacities for diapause: Comparative life-history traits. *Oecologia* 87: 146–151.

Spitze, K. 1991. *Chaoborus* predation and life-history evolution in *Daphnia pulex*: Temporal pattern of population diversity, fitness, and mean life history. *Evolution* 45: 82–92.

Spitze, K. 1992. Predator-mediated plasticity of prey life history and morphology: *Chaoborus americanus* predation on *Daphnia pulex*. *Amer. Nat.* 139: 229–247.

Stearns, S. C. 1976. Life-history tactics: A review of the ideas. *Quart. Rev. Biol.* 51: 3–47.

Stearns, S. C. 1977. The evolution of life history traits: A critique of the theory and a review of the data. *Ann. Rev. Ecol. Syst.* 8: 145–171.

Stearns, S. C. 1983a. The evolution of life history traits in mosquitofish since their introduction to Hawaii in 1905: Rates of evolution, heritabilities, and developmental plasticity. *Amer. Zool.* 23: 65–75.

Stearns, S. C. 1983b. The genetic basis of differences in life history traits among six populations of mosquitofish (*Gambusia affinis*) that shared ancestors in 1905. *Evolution* 37: 618–827.

Stearns, S. C., and J. C. Koella. 1986. The evolution of phenotypic plasticity in life-history traits: Predictions of reaction norms for age and size at maturity. *Evolution* 40: 893–913.

Stearns, S. C., and R. D. Sage. 1980. Maladaptation in a marginal population of mosquitofish, *Gambusia affinis*. *Evolution* 34: 65–75.

Stearns, S. C., G. de Jong, and R. A. Newman. 1991. The effects of phenotypic plasticity on genetic correlations. *Trends Ecol. Evol.* 6: 122–126.

Tauber, M. J., C. A. Tauber, and S. Masaki. 1986. *Seasonal Adaptations of Insects*. Oxford University Press, New York.

Taylor, D. R., and L. W. Aarssen. 1988. An interpretation of phenotypic plasticity in *Agropyron repens* (Gramineae). *Amer. J. Bot.* 75: 401–413.

Taylor, H. M., R. S. Gourley, C. E. Lawrence, and R. S. Kaplan. 1974. Natural selection of life history attributes: An analytical approach. *Theoret. Pop. Biol.* 5: 104–122.

Tinkle, D. W., and R. E. Ballinger. 1972. *Sceloporus undulatus*: A study of the intraspecific comparative demography of a lizard. *Ecology* 53: 27–34.

Tinkle, D. W., and A. E. Dunham. 1986. Comparative life histories of two syntopic sceloporine lizards. *Copeia* 1986: 1–18.

Tinkle, D. W., J. D. Congdon, and P. C. Rosen. 1981. Nesting frequency and success: Implications for the demography of painted turtles. *Ecology* 62: 1426–1432.

Townshend, T. J., and R. J. Wootten. 1984. Effects of food supply on the reproduction of the convict cichlid, *Cichlasoma nigrofasciatum*. *J. Fish Biol.* 24: 91–104.

Travis, J. 1989. Ecological genetics of life-history traits in poeciliid fishes. In *Ecology and Evolution of Livebearing Fishes (Poeciliidae)*, ed. G. K. Meffe and F. F. Snelson, Jr., pp. 185–200. Prentice Hall, New York.

Travis, J. In press. Evaluating the adaptive role of morphological plasticity. In *Ecomorphology: Integrative Organismal Biology*, ed. P. C. Wainwright and S. Reilly. University of Chicago Press.

Travis, J., and J. C. Trexler. 1987. Regional variation in habitat requirements of the sailfin molly, with special reference to the Florida Keys. *Nongame Wildlife Technical Report Number 3*, Florida Game and Fresh Water Fish Commission, Tallahassee.

Travis, J., J. A. Farr, S. Henrich, and R. T. Cheong. 1987. Testing theories of clutch overlap with the reproductive ecology of *Heterandria formosa*. *Ecology* 68: 611–623.

Trendall, J. T. 1982. Covariation of life history traits in the mosquitofish, *Gambusia affinis*. *Amer. Nat.* 119: 774–783.

Trendall, J. T. 1983. Life history variation among experimental populations of the mosquitofish, *Gambusia affinis*. *Copeia* 1983: 953–963.

Trexler, J. C. 1985. Variation in the degree of viviparity in the sailfin molly *Poecilia latipinna*. *Copeia* 1985: 999–1004.

Trexler, J. C., J. Travis, and M. Trexler. 1990. Phenotypic plasticity in the sailfin molly, *Poecilia latipinna* (Pisces: Poeciliidae). II. Laboratory experiment. *Evolution* 44: 157–167.

Turelli, M. 1988. Population genetic models for polygenic variation and evolution. In *Proceedings of the Second International Conference on Quantitative Genetics*, ed. B. S. Weir, E. J. Eisen, M. M. Goodman, and G. Namkoong, pp. 601–618. Sinauer, Sunderland, Mass.

Venable, D. L., and J. S. Brown. 1988. The selective interactions of dispersal, dormancy, and seed size as adaptations for reducing risk in variable environments. *Amer. Nat.* 131: 360–384.

Vetukhiv, M. 1957. Longevity of hybrids between geographic populations of *Drosophila pseudoobscura*. *Evolution* 11: 348–360.

Vitt, L. J., R. C. Van Loben Sels, and R. D. Ohmart. 1978. Life history traits of lizards and snakes. *Amer. Nat.* 125: 480–484.

Wilbur, H. M. 1976. Life history evolution of seven milkweeds of the genus *Asclepias*. *J. Ecol.* 64: 223–240.

Wilbur, H. M. 1977. Propagule size, number, and dispersion pattern in *Ambystoma* and *Asclepias*. *Am. Nat.* 111: 43–68.

Wilbur, H. M., and J. P. Collins. 1973. Ecological aspects of amphibian metamorphosis. *Science* 182: 1305–1314.

Wilbur, H. M., and P. J. Morin. 1988. Life history evolution in turtles. In *Biology of the Reptilia*, ed. C. Gans and R. B. Huey, vol. 16B, pp. 387–439. Liss, New York.

Wilbur, H. M., D. W. Tinkle, and J. P. Collins. 1974. Environmental certainty, trophic level, and resource availability in life history evolution. *Amer. Nat.* 108: 805–817.

Wilken, D. H. 1977. Local differentiation for phenotypic plasticity in the wild California annual *Collomia linearis* (Polemoniaceae). *Syst. Bot.* 2: 99–108.

Winkler, D. W., and K. Wallin. 1987. Offspring size and number: A life history model linking effort per offspring and total effort. *Amer. Nat.* 129: 708–720.

Winn, A. A. 1985. Effects of seed size and microsite on seedling emergence of *Prunella vulgaris* in four habitats. *J. Ecol.* 73: 831–840.

Winn, A. A. 1988. Ecological and evolutionary consequences of seed size in *Prunella vulgaris*. *Ecology* 69: 1537–1544.

Winn, A. A., and A. S. Evans. 1991. Variation among populations of *Prunella vulgaris* in plastic responses to light. *Functional Ecol.* 5: 562–571.

Winn, A. A., and P. A. Werner. 1987. Regulation of seed yield within and among populations of *Prunella vulgaris*. *Ecology* 68: 1224–1233.

Wootten, R. J. 1973. The effect of size of food ration on egg production in the female threespined stickleback. *J. Fish Biol.* 5: 89–96.

Wourms, J. P., B. D. Grove, and J. Lombardi. 1988. The maternal-embryonic relationship in viviparous fishes. In *Fish Physiology*, ed. W. S. Hoar and D. J. Randall, vol. 11B, pp. 1–134. Academic Press, New York.

Wulff, R. D. 1986a. Seed size variation in *Desmodium paniculatum*. I. Factors affecting seed size. *J. Ecol.* 74: 87–97.

Wulff, R. D. 1986b. Seed size variation in *Desmodium paniculatum*. II. Effects on seedling growth and physiological performance. *J. Ecol.* 74: 99–114.

Wyngaard, G. A. 1986a. Genetic differentiation of life history traits in populations of *Mesocyclops edax* (Crustacea: Copepoda). *Biol. Bull.* 170: 279–295.

Wyngaard, G. A. 1986b. Heritable life history variation in widely separated populations of *Mesocyclops edax* (Crustacea: Copepoda). *Biol. Bull.* 170: 296–304.

Zammuto, R. M., and J. S. Millar. 1985a. A consideration of bet-hedging in *Spermophilus columbianus*. *J. Mamm.* 66: 652–660.

Zammuto, R. M., and J. S. Millar. 1985b. Environmental predictability, variability, and *Spermophilus columbianus* life history over an environmental gradient. *Ecology* 66: 1784–1794.

Zangerl, A. R., and M. R. Berenbaum. 1990. Furanocoumarin induction in wild parsnip: Genetics and populational variation. *Ecology* 71: 1933–1940.

Zeng, Z.-B. 1988. Long-term correlated response, interpopulation covariation, and interspecific allometry. *Evolution* 42: 363–374.

10

Evolution in the Sailfin Molly: The Interplay of Life-History Variation and Sexual Selection

Introduction

Evolution can be defined as a process of descent with modification, a process that converts genetic variability within species into genetic variation among species. In the initial stages of this process, genetic variability within populations is converted into genetic variation among conspecific populations. This conversion occurs through some combination of natural selection and a variety of "random" processes that constrain how effectively selection can produce differentiation.

These statements describe our current vision of evolution in an accurate but soulless fashion. The soullessness emerges for two reasons. First, general terms like "natural selection" and "'random' processes" cover a multitude of interesting biological and historical phenomena. Moreover, vague phrases like "genetic variation among populations" do not connote any sense of how much differentiation can occur (or how little does occur) and how the genetic variation for different traits in a species is structured into components within and among populations.

Second, and more critically, these descriptive statements hide our ignorance about the relative importance of the various forces that drive the process of differentiation. Questions about "relative importance" appear in various guises, as pertaining to the efficacy of selection in the face of gene flow, the importance of local selection in the face of fluctuating population sizes, the availability of suitable genetic variation for one key trait in the face of conflicting selective pressures on many traits, and the importance of local genetic variation from which adaptations might be fashioned in the face of the chance events of dispersal and local extinctions. Although we have an abundance of excellent theory for population differentiation and a core of solid empirical studies of differentiation in individual traits, these questions remain open empirical ones that await a collection of comprehensive examinations of general intraspecific variation.

Comprehensive studies are important for two reasons. First, organisms are complex entities, and local populations face an array of abiotic and biotic challenges. Local adaptation, if it occurs, will inevitably involve many traits and necessitate compromises among trait values. Thus, problems of physiology will impinge on those of life history, which will affect those of behavior, which in turn may be reflected in physiology; although biologists often specialize in subdisciplines, their objects of study do not, and only comprehensive studies can illuminate the complexity of the evolutionary process. Second, the historical context of differentiation will vary from species to species. We will gain some sense of how important the role of history and its concomitant constraints on evolution are only from a series of comprehensive studies of intraspecific variation.

The thesis of this paper is, therefore, that such comprehensive studies offer critical insights into the evolutionary process. I will illustrate this thesis with a summary of the studies by my colleagues and myself (and others) on morphological, behavioral, and life-history variation in the sailfin molly, *Poecilia latipinna*. In the next section I offer an overview of theoretical and empirical literature on population differentiation as a context for our studies of the sailfin molly. In the subsequent sections I review the patterns of differentiation in sailfin mollies and what is known of the causes of that differentiation. In the final section I discuss the general lessons that the study of this system has offered.

Molding Genetic Variation: Patterns of Polymorphism and Local Differentiation

The Conceptual Complexities of Adaptive Differentiation

The initial stages of adaptive differentiation among nonisolated populations are usually envisioned to occur as shifts in the relative frequencies of alternative alleles as a result of locally varying selection pressures (Nagylaki 1975; Nagylaki and Lucier 1980). Within each constituent population varying levels of polymorphism are maintained by overdominance and gene flow–selection balance (Nagylaki 1975; Endler 1977). The best exemplars of this paradigm are patterns of polymorphism and differentiation for certain allozymes, in which local differences in allele frequencies are generated by local differences in fitness in response to abiotic factors (Koehn et al. 1983; Powers et al. 1983, 1986; Chappell and Snyder 1984). Studies of local or geographic variation in discrete morphological characters have also been fit into this pattern (Endler 1983; Borowsky 1984).

Patterns of polymorphism and differentiation in continuous phenotypic characters such as life-history traits can be generated in analogous fashion. Local populations may exhibit variation in the optimal value of the trait, while within each population weak optimizing selection and gene flow–selection balance maintain genetic variation (Slatkin 1975, 1978). Differences among populations may be graded, or populations may display sharp differences depending on the relative strength and spatial scales of the forces of gene flow and selection (Slatkin 1973, 1978; Tachida and Cockerham 1987).

This caricature of differentiation in continuous traits is deceptively simple because it describes general patterns of *genotypic* variation under weak local selection. The observed spatial patterns of genotypic and phenotypic variance in the *trait* will be affected in complicated ways by several additional factors. Different modes of gene action can produce different patterns of among- and within-population genetic variances (Tachida and Cockerham 1987). The genetic control of such traits almost always involves pleiotropic effects (Lynch 1985), which implies that differentiation in one character cannot be understood without consideration of the constraints imposed by local selection on genetically correlated characters. This effect can introduce a host of complications; a correlated character may be under selection in only some locations (Zimmerer and Kallman 1988), and epistatic modifiers of the primary genes may be distributed differentially among locations (Lande 1983; Kallman 1989) but play major roles in affecting fitness (MacDonald et al. 1978; Matsuo and Yamazaki 1984), phenotypic variance (Laurie-Ahlberg et al. 1981), and genetic correlations (Wilton et al. 1982; Lenski 1988) in all locations. Metric characters are affected by environmental factors that may operate at a different spatial scale than that at which selection and gene flow operate (i.e., the characteristic length: Slatkin 1973), rendering interpretation of patterns of trait differentiation ambiguous without experimentation (Berven et al. 1979; James 1983; Bernays 1986). Genotype-specific migration patterns may obscure the detection of trait differentiation and the spatial scale at which local selection occurs (Moody 1981). Finally, strong selection can readily overcome gene flow and generate local differentiation for primary traits. However, in many cases, selection may be acting on a correlative pattern between two traits (Bürger 1986; Zeng 1988) or on the way in which the developmental system expresses the trait in response to commonly encountered environmental conditions (Policansky 1983; Stearns and Koella 1986). In these cases selection is "second order," not acting on a particular genotype but on an interaction of genotype with environment or of genotype at one locus with genotype at another locus. Studies in a variety of contexts have found that precise differentiation in second-order traits is very difficult in the face of even low levels of gene flow (Caisse and Antonovics 1978; Via and Lande 1985; Zeng 1988).

Ultimate patterns of local differentiation in metric traits can also be affected profoundly by the mode of differentiation. Many models of differentiation are caricatures of gradual divergence of descendant populations from a common ancestral population. However, divergence of local populations may occur through a process more like "dispersal" than "vicariance." A new habitat is colonized by a founder group of small size, the new population expands rapidly, a new set of colonists is disgorged into an adjacent area, and the cycle repeats. This "dispersal" process can generate more variety in the consequent patterns of differentiation than the "vicariant" process because of two effects. First, genetic drift in the colonization process will affect direct and correlated responses to selection (Eisen et al. 1973). Second, the ultimate responses to selection in finite, inbreeding populations involve additive, dominance, and epistatic genetic variances in idiosyncratic combinations (Cockerham and Tachida 1988; Goodnight 1988).

Empirical Evidence

The study of stable local patterns of differentiation for metric traits under moderate–strong selection is an important topic. First, the problem is central to current debates about the importance of evolutionary processes within species in generating diversity and the actual amount of genetic change on which adaptive diversity is based (Coyne and Beecham 1987; Mettler et al. 1988; Stanley 1988). Second, most models of differentiation employ weak selection; studies of traits under strong selection can reveal the extent to which those models nonetheless provide a sufficient qualitative description of the microevolutionary process.

Perusal of the empirical literature on animals reveals several gaps in our current knowledge. First, most studies of the ecology and genetics of adaptive differentiation have examined populations that experience an unknown but presumably low level of reciprocal gene exchange (Hirshfield et al. 1980; Berven 1982; Reznick 1982; Stearns 1983; Lonsdale and Levinton 1985; Breden et al. 1987). Studies of differentiation in the face of gene flow have focused on a subset of systems: allozymes within a specialized breeding structure (Williams and Koehn 1984), localized anthropogenous effects (Antonovics et al. 1971), host races of herbivorous insects (Futuyma 1983; Via, chap. 4, this volume), or situations of secondary contact (Powers et al. 1986). The lack of attention to a wide range of systems means that the empirical importance of local selection, local environmental effects, and gene flow in determining the spatial scale of differentiation remains debatable (Ball et al. 1988). Correlations between graded differences among populations in allele frequencies or trait values and graded differences in environmental variables have been traditional lines of circumstantial evidence in favor of fine-scale adaptation (Ballinger 1979; Nevo and Yang 1979), but the potential for gene flow and environmental effects to produce such gradations weakens the inference that can be made from correlations alone (James 1970).

Second, the spatial scale at which adaptation occurs for different types of traits is unresolved. Local differentiation for second-order traits is not a well-documented pattern, although adaptive differentiation for such traits at a geographic scale has been demonstrated (Bradshaw 1986; Conover and Heins 1987; Coyne and Beecham 1987). Is the lack of local variation due to a lack of sufficiently localized selection pressures, a lack of the requisite genetic variance, or the irresistible effect of gene flow?

Third, the effects of differentiation on genetic parameters within populations are not well understood. Changes in intrademic genetic parameters may (Stearns 1984; Berven 1987) or may not (Ayres and Arnold 1983) accompany adaptive differentiation, but the reasons for this difference in outcome are unknown (see Lynch et al., chap. 6, this volume).

These gaps illustrate the need for comprehensive studies of the ecology and genetics of an ensemble of local populations that appear to display stable patterns of adaptive differentiation despite ongoing gene exchange. The system should be amenable to comparative ecological and genetic analysis. Our work on the sailfin molly has developed into such a systematic study.

Biology of the Sailfin Molly

Patterns of Variation

The sailfin molly is a common fish of salt marshes, brackish impoundments, and specialized freshwater habitats throughout the southern Atlantic and Gulf coastal areas (Lee et al. 1980). A member of the topminnow or livebearer family (Poeciliidae), the sailfin molly has internal fertilization, and females hold developing embryos until the young can live independently. Energy for embryonic development is based in part on yolk in the ovum and in part on nutrient transfer to developing embryos (= viviparity; Trexler 1985).

Studies of allozyme variation have revealed substantial amounts of genetic variation in all populations (Simanek 1978; Trexler 1988). From 14% to 24% of loci have been found to be polymorphic with two to five alleles at such loci. The average heterozygosity levels per locus ranged from 5% to 9% among populations, which are typical values for teleost fish. Preliminary surveys of mtDNA variation have supported the conclusion that molly populations harbor healthy levels of genetic variation (Avise et al. 1991; Ptacek and Travis, unpublished data). There is little indication from such levels of genetic variation that small population

TABLE 10.1

Variance component ratios for allozyme variation, male body size, and body size of gravid female mollies from the southeastern United States

Factor	Allozymes	Male Size	Female Size
Among regions	0.22	0.06	0.07
Among populations within regions	0.03	0.25	0.18
Within populations	0.75	0.69	0.75

Notes: Entries are the ratios of the variance component for that factor relative to the total variance. Allozyme parameters are based on twenty-six loci and are taken from Trexler 1988; body size parameters are drawn from Travis and Trexler 1987, which should be consulted for details.

processes have played any lasting significant role in the diversification of mollies in this part of the species range.

Mollies display a simple pattern of "isolation by distance" (Simanek 1978; Trexler 1988). In the eastern half of the species' range, differentiation at allozyme markers is gradual, although there are sharp differences across the Apalachicola and Mississippi rivers (Simanek 1978). In the coastal-plain populations of Georgia and Florida, there are few "private alleles," and coancestry statistics show that local differentiation accounts for only 5% to 12% of variation within a region (table 10.1). Variation among geographic regions accounts for 22% of all variation. Repeated surveys (Simanek 1978; Trexler 1988) uncovered little change in heterozygosity and allele frequencies across sampling periods, suggesting that mollies have large effective population sizes and levels of effective migration that are comparable to the highest rates estimated in animals (Trexler 1988; Slatkin, chap. 1, this volume). These conclusions match expectations from the natural history (see Trexler 1988; see Lynch et al., chap. 6, this volume, for another example). However, these patterns are very different from those in the related genus *Xiphophorus* (Borowsky 1984) and other poeciliids (Smith et al. 1989).

Spatial autocorrelation analyses offer more precise insights into the scale of allozyme differentiation (Trexler 1988). Demes separated by less than 5 km have very similar allele frequencies; demes separated by 6 to 51 km are less similar, on average, than closer pairs of demes, but still show similar allele frequencies. When two demes are separated by 51 to 200 km, their allele frequencies are uncorrelated, and demes separated by more than 200 km have divergent allele frequencies. These patterns suggest that populations within a 5-km distance will be closely interconnected genetically (Antonovics et al., chap 8, this volume) and that significant differentiation for nearly-neutral alleles requires populations to be separated by at least 51 km. Thus the expected spatial scale of neutral local differentiation should be rather large.

Trexler's (1988) survey found a trend toward more local differentiation among northern groups of demes (Georgia) than among southern groups (Everglades and Florida Keys; Trexler 1988). This pattern indicates either different equilibrial levels of gene flow among local demes in the different areas or different temporary historical displacements from a common equilibrial configuration for the species (see Slatkin, chap. 1, this volume). Different displacements from such an equilibrium from north to south are compatible with a historical explanation based on recent zoogeography and current climatic regimes. More northern populations have probably been extant longer and appear less subject to either the periodic extinctions experienced by Everglades populations as a result of extreme drought and wading bird predation (Trexler, pers. obs.) or the periodic disruptions of population structure experienced by Keys populations as a result of frequent tropical storms and hurricanes.

Mollies are notable for the extensive phenotypic variation in male body size seen within populations throughout their range (Hubbs 1942, 1964; Kilby 1955; Simanek 1978; Snelson 1985). Growth in males is negligible after maturity (Snelson 1982, 1985; Travis et al. 1989), so variation in male size is a direct reflection of variation in size at maturity. Individual populations have characteristic distributions of male body size that are constant across dozens of generations (Travis and Trexler 1987; Travis 1989). These stable differences among populations in average male size and their characteristic phenotypic variances in male size are *completely unrelated* to geographic proximity, unlike allozyme differences (Travis and Trexler 1987). The magnitude of the difference can be large; the average values seen in closely adjacent (< 10 km) populations typically differ by 20% to 50% (Travis and Trexler 1987; Trexler et al. 1990).

The body sizes of gravid females exhibit patterns that parallel the patterns in males. There is extensive variation within populations, but unlike the variation among males, variation among females is attributable almost entirely to age structure (Trexler, unpublished data). Variation in the average female body size in the eastern half of the species' range displays a hierarchical pattern similar to that of males (table 10.1), and in fact the average body size of gravid females in a population exhibits a strong correlation (from 0.63 to 0.70) with that of males (fig. 10.1).

Body-size variation appears to be the major axis along which female life history varies. Unlike patterns in other poeciliids (Borowsky and Diffley 1981), there is little variation anywhere in the eastern half of the species' range in the slope of the relationship between the log of embryo

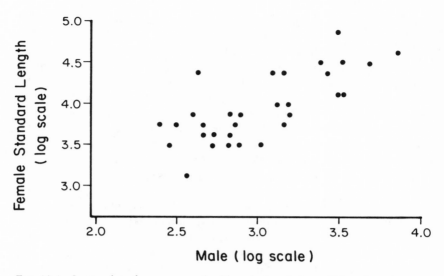

FIG. 10.1. Scatterplot of average standard lengths (mm) of gravid female and mature male mollies from samples taken at populations in the coastal plain of the southeastern United States in 1986. Each point represents an average based on at least 10 females and 20 males per population. (Data from Trexler 1986.) A similar pattern is seen in data from the same populations taken in 1985.

TABLE 10.2
Male and female body lengths, reproductive season, and fecundity at several north Florida populations (data taken 1983–1987)

Location	Description (Salinity)	Season[a]	Average Female Size (mm)	Average Number of Offspring per Brood	Average Adjusted Brood Size[b]	Average Male Size (mm)
Wacissa	Freshwater	4	30.0	15	25	25.8
Mounds	Brackish (1–5‰)	4	36.1	17	20	24.2
Lighthouse	Brackish (4–8‰)	5	36.3	20	19	23.6
Boat Ramp	Tidal pool (8–15‰)	5	40.4	28	23	27.8
Live Oak	Tidal stream (10–20‰)	6	44.9	48	28	32.4
Melanie's	Tidal pool (15–22‰)	7	45.2	35	31	37.0

[a] Number of continuous months (March–September) in which gravid females can reliably be collected.

[b] Mean values of number of offsprings per brood adjusted for differences in female body size among locations via analysis of covariance.

number and the log of female standard length (Travis and Trexler 1987). There are some consistent differences among populations on the local scale in north Florida. In more brackish-water and freshwater populations, average female size is smaller, there are fewer embryos for a female of a given size, the duration of the reproductive season is shorter, and the average size of *males* is smaller (table 10.2). The degree of viviparity appears to be a phenotypically plastic trait (Trexler 1985). There is no relationship between the heterozygosity of a female and either her standard length or her average embryo number (adjusted for body size) within any of several populations (Travis 1989).

The sailfin molly has a pronounced sexual dimorphism in the length and height of the dorsal fin, with considerably exaggerated heights in many males (Luckner 1979; Snelson 1985). The size of the fin increases in positively allometric fashion to male standard length, the degree of allometry is the same across populations, but there is an interesting "countergradient" phenomenon first noted by C. L. Hubbs (1942): dorsal fins are disproportionately higher in populations in which the average male size is consistently smaller (Farr et al. 1986).

Male mollies attempt to obtain matings through two behavioral interactions with females. Courtship displays are active attempts to elicit female interest and cooperation. Forced insemination attempts (gonopodial thrusts) are attempts to inseminate females without their cooperation. The behavioral profile of a male varies as a function of his size; the rates of courtship displays increase allometrically and those of forced insemination attempts decrease allometrically with increases in male size (Farr et al. 1986; fig. 10.2). In the extreme, large males (ca. 50 mm standard length) continuously court females, whereas small males (ca. 25 mm standard length) never engage in courtship and attempt to obtain all matings through forced insemination attempts. These behavioral patterns are highly repeatable among individuals (Travis and Woodward 1989) and are seen in uncontrolled field observations (table 10.3) as well as controlled laboratory studies.

Behavioral profiles vary among molly populations. First, because of the size dependence of these profiles, populations in which males are small, on average, are populations in which courtship displays are uncommon. However, there is a surprising amount of behavioral variation among populations even when the differences in male size have been accounted for by analysis of covariance (Farr et al. 1986). This variation in "adjusted" behavior rates has a particular pattern. Variation among populations in behavioral profiles is based first on variation in body size but secondarily, as noted above, on variation in rates of behavior that is independent of size (Farr et al. 1986). The same "countergradient" pattern that occurs in dorsal-fin height occurs even more strongly in the rate of

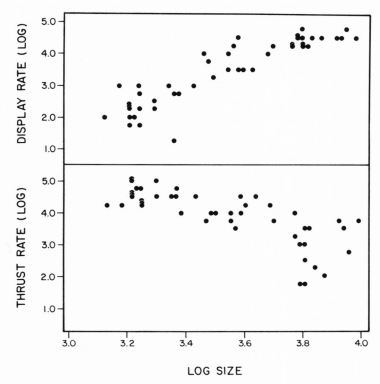

FIG. 10.2. Scatterplots (logarithmic scales) of the number of courtship displays (*top*) and forced insemination attempts (gonopodial thrusts; *bottom*) exhibited by individual male mollies in a 30-minute observation period under laboratory conditions as a function of the standard length of the male. Males were the progeny of controlled crosses and were raised individually in the laboratory. (Data from Travis et al., in press [a])

TABLE 10.3
Rates of behavior exhibited by males of different length classes in natural populations

	Male Length Class		
Category	*Small*	*Medium*	*Large*
Range of lengths (mm)	17–25	30–40	45–55
Number of observations	69	67	128
Gonoporal nibbles	3.6 ± 0.4	3.1 ± 0.5	1.4 ± 0.2
Gonopodial thrusts	2.0 ± 0.4	1.0 ± 0.2	0.2 ± 0.1
Courtship displays	0.1 ± 0.0	1.3 ± 0.4	2.8 ± 0.3
Total observed behaviors	5.6 ± 0.7	5.5 ± 0.7	4.4 ± 0.3

Note: Entry for each behavior is the average and standard error of the number of times that behavior was observed for a focal individual in a one-minute observation period.

courtship displays. The spreading of an erect dorsal fin plays a prominent role in that display. No such "countergradient" effects are seen in dorsal-fin length or in the rate of forced insemination attempts (Farr et al. 1986).

Genetics of Body Size and Life-History Traits

Variation in male body size at maturity is inherited patriclinally (Travis et al., in press [b]; fig. 10.3). Patriclinal inheritance of body size is known for many poeciliids (Kallman 1989), and much of the pattern of inheritance in mollies is shared by species of *Xiphophorus* in which males have XY sex chromosomes. Males with allelic variants on the Y chromosome have similar juvenile growth rates but initiate sexual differentiation at different ages, producing a strong genetic correlation between the age and size at maturity. A small (ca. 25 mm SL) male will complete maturation in about seventy days, whereas a large (ca. 50 mm SL) male will take about two hundred days. There is evidence that sex-limited autosomal factors modify the final size at a given age. When fish from different populations are raised in a common environment, there are significant differences among

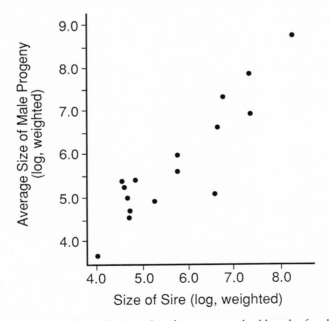

FIG. 10.3. Scatterplot (logarithmic scale) of average standard length of male progeny in a paternal half-sib family and the standard length of their father. Each point is the average of 2–12 males and is weighted by the family size. (Data from Travis et al., in press [b])

the male progeny from different populations in the average ages and sizes at maturity (Trexler 1986; Trexler et al. 1990).

There is little variation in female size or age at maturity within a population (Travis et al., in press [b]), although small but significant differences in the age at maturity are seen among females from different populations when fish are raised in a constant environment (Trexler 1986, 1989). There are no genetic correlations between genders for age or size at maturity. Female body-size variation within natural populations appears primarily due to age-class effects (Trexler, unpublished otolith data).

Secondary sex characters exhibit more complex sources of variance. Most of the variance in such characters is generated by covariances with male body size, but there are large, significant genetic effects on rate of courtship displays that are independent of the inheritance of body size (Travis et al., in press [a]). No such effects were seen for the rates of gonoporal nibbles or gonopodial thrusting. Dorsal fin length and height are almost completely covarying with body size; there are small genetic effects on the height of the fin that occur independently of the inheritance of size. These effects are uncorrelated with whatever factors modify the expression of behavior for a given body size. Social experience during ontogeny has no appreciable effect on a male's size-specific behavior pattern (Farr and Travis 1989), and there are only minor effects of age (Travis et al., in press [a]).

Environmental Sources of Variation and Genetics-Environment Interactions

Males and females respond differently to environmental effects (Trexler and Travis 1990; Trexler et al. 1990). Juvenile growth rate of females was much more sensitive than that of males to the depressant effects of low temperatures and low salinities. At cooler temperatures and lower salinities, males delay maturation moderately and eventually mature at slightly smaller sizes than they achieve at warmer temperatures and higher salinities. By contrast, females delay maturation dramatically at cooler temperatures and lower salinities, but they mature at a comparable size across all conditions.

There is no evidence for any ontogenetic behavioral effects that would alter male behavior profiles at a given male body size (Farr and Travis 1989). When males are stunted (relative to half-full siblings with the same Y chromosome), they display patterns of behavior more in accord with their phenotype than with their genotype. The patterns of behavior for a given body size appear to have a largely ontogenetic origin (Travis et al., in press [a]).

Polymorphism and Differentiation in the Sailfin Molly

An obvious hypothesis for the maintenance of size variation within molly populations invokes a balance between a demographic component of fitness that might favor the smaller males and the component of fitness involving mating and reproductive success that might favor the larger males. Local variation in size patterns would be generated by local variation in the factors that set this balance. In this section, I review the evidence in favor of this hypothesis.

Natural Selection

The threefold difference in age at maturity between males at the extremes of the size distribution generates a substantial fertility differential because there can be extensive overlap of generations (Charlesworth 1980). The effect can be best appreciated from the fact that, of a small and a large male born at the same time, the large male will be recruited into the adult segment of the population at the same time as the sons of his smaller contemporary. The differential is worsened at cooler temperatures and lower salinities by two factors: the time to achieve maturity is longer for all males, and the actual disparity among genotypes is increased (Trexler 1986, 1989).

These results implicate the scope for growth (Brett 1979) as a major factor in determining fertility differences among male genotypes. Environmental effects that increase the maintenance demand or that lower the available energy increase the time to maturity and thereby put larger males at a greater disadvantage. The importance of this effect will be determined by whether the reproductive season is long enough to allow the large males an opportunity to redress the imbalance. Duration of reproduction is a function of the energetic regime because females respond to lowered scopes for growth due to salinity and temperature by growing more slowly, taking longer to mature, and having longer gestation periods (Trexler 1986, 1989; Snelson et al. 1986). Where energetic relationships permit longer reproductive seasons and higher fecundities, larger males have the opportunity to redress the imbalance wrought by their longer maturation times through enhanced reproductive success.

The correlation of male body-size variation with that of gravid females at either a geographic (fig. 10.1) or a local (table 10.2) scale offers indirect evidence in support of this idea. The correlation is striking because body-size variation in males is largely based on genetic variation, whereas that among females is largely environmental in origin (age or size at a given age). In less productive habitats or more energetically demanding ones, average female size is smaller, there are fewer embryos for a female of a given size, the duration of the reproductive season is shorter, and the average size of males is smaller. This local pattern is repeated throughout the eastern half of the species' range (Travis and Trexler 1987).

Salinity per se is unlikely to be the only controlling factor. A number of habitat-specific factors are correlated with and include salinity, and the combination will affect the scope for growth and the scope for reproduction in these populations and thereby govern the body-size variation. There is evidence to support this contention. Experiments on osmoregulation and growth patterns suggest that estuarine populations perform best at ambient osmolalities around the isosmotic point (ca. 360 mOsm/kg), although there is little difference through a range from about 150 mOsm/kg to 600 mOsm/kg. Individuals have higher respiration rates, less precise ionic regulation, lower growth rates, longer times to maturity, and generally lower condition on either side of that range (Evans 1973, 1975, 1984; Gustafson 1981; Zimmerer 1983; Trexler 1986, 1989).

However, stressful abiotic conditions encountered after maturity may impose viability selection that favors larger fish. We have found that larger fish are often, but not invariably, more likely to survive winter conditions in low-density field enclosures (Trexler et al. 1992). Observations from natural populations suggest that this result is due to energetic stress. Analysis of covariance of male sizes in the nonbreeding, nongrowing season shows that somatic dry mass for a given standard length (grand mean of 32.8 mm) decreases from an average of 276 mg in October to 237 mg in December, slowly increases to 246 mg (January), and is up to 255 mg in the beginning of the reproductive season (March). These differences are significant ($F_{(3,72)} = 4.63$, $P < 0.005$, 80 males total) and indicate a loss of condition over the winter. Similar patterns have been reported in another estuarine fish (Conover and Ross 1982; Conover 1984). Subsequent experimental work by M. McManus (unpublished data) has confirmed this hypothesis: larger males have greater stores of triacylglycerol and these stores are exploited for survival under winter conditions.

The advantage of larger size will be minimized in some habitats because of predation by wading birds. Wading birds preferentially prey on larger fish (Britton and Moser 1982; Trexler et al., in press), but this effect will be important only in some locations. Wading birds forage less effectively in water with less clarity (Kushlan 1978), greater vegetation structure (Custer and Osborn 1978), and greater depth (Kramer et al. 1983). In murky water atop soft muddy bottoms, as in many north Florida salt marshes, bird predation will not be an effective force governing size structure. By contrast, in the shallow, clear water atop the hard substrates of the Everglades, wading-bird predation is probably a major reason for the small body size of mollies in that location (Trexler et al., in press).

Population Structure and Mating Success

BEHAVIORAL EVIDENCE

In the field, most females receive no attention from a male in a 60-second observation period, whereas a few females receive attention from up to six different males in the same period (table 10.4). Among those females who receive any attention from a male, larger females receive more attention. Individual males were always observed to approach at least one female in 60 seconds, but they gave persistent attention to only a few of those originally approached (table 10.5). Larger males were likely to persist in devoting attention to a single female, once that female was approached and "chosen"; smaller males moved from female to female quite readily.

TABLE 10.4
The number of males that approach a focal female molly of different size classes in a one-minute observation period in the field

Size Class	Average ± se	Patchiness ± se	Sig?	N
Small	0.83 ± 0.15	2.03 ± 0.30	NS	66
Medium	1.38 ± 0.14	1.67 ± 0.05	Yes	129
Large	1.63 ± 0.17	1.68 ± 0.05	Yes	113

Note: The first column is the average (± one standard error) number of males, while the second column is the value of Lloyd's (1967) patchiness index; values above one indicate a clumped distribution of males per female, while values below one indicate a regular distribution. Clumped distributions indicate that a few females were approached by many males but that most females were not approached at all. Ranges of sizes in each size-class correspond to those of the males in table 10.3.

TABLE 10.5
The probabilities (with standard errors) that a male molly will persist in giving attention to a female (= follow-up), once he has approached her, for females in different size classes in a one-minute observation period in the field

Size Class	Probability (Follow-up/Approach)	Average Number, Given Approach
Small (32)	0.25 ± 0.08	0.37 ± 0.15
Medium (78)	0.50 ± 0.06	0.74 ± 0.11
Large (72)	0.69 ± 0.05	1.10 ± 0.13

Note: All distributions are random. The last column is the average number of males who persisted with a given female, given that each male approached her. Numbers in parentheses indicate sample size of observation periods for each female size class (ranges of sizes are as in tables 10.3 and 10.4).

Laboratory studies reveal that females have cycles of ovarian receptivity and are capable of fertilizing a group of ova once every twenty-eight to thirty-two days. Females do not cycle synchronously, and on any single day only a few females will be in receptive condition. Gravid, nonreceptive females do not respond to courtship displays, and they sometimes behave aggressively toward males who attempt forced inseminations. Receptive females respond cooperatively to courtship displays, but when given a choice of males, they are drawn toward larger males. Males recognize receptive females, and large males preferentially devote attention to them. Small males will also devote more attention to a female with whom they are familiar when she becomes receptive, but, unlike large males, they do not discriminate between a receptive and a nonreceptive female if she is novel and they are given a brief exposure. Large males aggressively dominate small males and limit small males' access to receptive females. When given a choice between receptive females of different sizes, males preferentially direct their attention to the larger females (synthesized from Hubbs 1964; Baird 1968, 1974; Parzefall 1969, 1973, 1979; Luckner 1979; Balsano et al. 1981; Woodhead and Armstrong 1985; Farr and Travis 1986; Farr et al. 1986; Travis and Woodward 1989; Sumner et al., in press; unpublished data of J. Travis).

The laboratory observations are consistent with the patterns of association seen in the field. Only a few females are objects of male attention on any given day, and they are presumably the receptive females. Larger males appear more persistent with these few females, and smaller males appear to be searching for any female they can have access to. Larger females in the field were more likely to be objects of investigation than smaller females, which matches the predilection for larger females seen in the laboratory.

HISTOLOGICAL EVIDENCE

Male mollies are active in sexual behavior and have constant levels of sperm production year-round, although more sperm is found retained in males collected in the nonbreeding season (Monaco et al. 1981). In contrast, reproduction in females is induced by the combination of long days and warm temperatures (Grier 1973). Sperm can be stored for long periods in all female poeciliids in crevices between epithelial cells lining the lumen of the ovary (Nagahama 1983). New sperm arriving when a female is receptive will fill the lumen and have a competitive edge over older stored sperm (Grier 1973). These results suggest that sperm competition will be most likely to be "won" by the sperm received during the receptive period. This sperm appears more likely than not to be that received from larger males.

ALLOZYME EVIDENCE

In a survey of twenty-three wild-caught females using fourteen electrophoretic loci, eleven females were demonstrably multiply inseminated, and twelve females were not (Travis et al. 1990). The females that were demonstrably multiply inseminated were longer (48.6 mm vs. 41.7 mm), had more embryos per brood (64.8 vs. 34.2), and had more embryos for a given female size (at the grand mean of 44 mm, 49 embryos vs. 33, $F_{[1,19]} = 8.20$, $P < 0.01$). These results suggest that larger, more fecund females may be the objects of intermale competition. If larger males preferentially sired broods of larger females, the larger broods due to female size variation alone would aid in the maintenance of the polymorphism by dramatically increasing the reproductive success of large males above what could be achieved by mating at random with respect to female size.

There is allozyme evidence compatible with assortative mating. Trexler's (1988) electrophoretic survey indicated that heterozygote deficiency accounted for about 8% of the genetic variation within populations. For some loci the magnitude of the effect was considerable (value of F(ID) was 0.194 for PGM-3), and two north Florida populations, Boat Ramp and Lighthouse Pond, displayed large deficiencies from Hardy-Weinberg proportions at this locus. The large effective size of these populations would seem to rule out inbreeding.

A final line of allozyme evidence confirms Simanek's (1978) claim that the level of genetic variation in a population is inversely related to the frequency of "large" males, presumably because large males obtain a disproportionate share of the matings. In north Florida populations the average heterozygosity displayed a negative partial correlation with the average body size of males in a population (–0.72) when the level of phenotypic variance in body size was held constant, and heterozygosity went up with the phenotypic variance (0.84) when the average was held constant (data from Trexler 1986, 1988).

Synthesis

Molly populations live in estuarine environments that vary temporally but that also display continuous variation among sites in a variety of important abiotic parameters. Abiotic factors play major roles as selective agents that affect the equilibrium proportions of the different male body size "morphs." Continuous variation in these factors should thus generate continuous variation in the net balance between natural and sexual selection, which in turn should generate continuous variation in male body size among populations. This is the actual pattern (Travis and Trexler 1987). This line of reasoning would predict considerable fine-scale differentiation, because many of these populations are closely adjacent. Indeed, because male body-size variation appears primarily to be caused by Y-linked allelic series, the spatial autocorrelation analyses of allozyme variation could be used as the "null hypothesis" for the expected spatial scale of body-size differentiation. In this light, the sharp differentiation in male body size among populations less than 10 km apart would seem to require an adaptive explanation.

There are alternative interpretations to consider. Continuous variation among populations could be generated by sharp genetic differentiation that is obscured by environmental effects on growth and development that operate at a different spatial scale. However, Trexler (1986, 1989; Trexler et al. 1990) and others have shown that the amount of plasticity in male size is too limited to explain the magnitude of interdemic variation on the basis of environmental effects alone. Moreover, there is no evidence for growth rate, age, or size at maturity that local populations differ in their norms of reaction to temperature, food level, and salinity (Trexler 1986, 1989; Trexler et al. 1990). This result is notable because there are genotype-environment interactions *within* populations (Trexler and Travis 1990) that should provide the raw material for divergent norms of reaction (Via and Lande 1985; Via, chap. 3, this volume).

High rates of gene flow could also contribute to "smooth" differences among populations, and patterns of allozyme differentiation in mollies suggest that high rates of gene flow occur. High gene flow would have several effects; it would increase variation within populations if the migrants come from populations with very different male sizes (Slatkin 1978; Tachida and Cockerham 1987); it would hinder the evolution of local variation in norms of reaction (Via and Lande 1985); and it could ensure similar patterns of correlations and allometry among traits across populations (Zeng 1988; Lynch et al., chap. 6, this volume). The data we have presented are compatible with all of these expectations, suggesting that ongoing gene flow plays a major role in molding patterns of local variation in mollies by smoothing first-order differentiation and preventing differentiation in second-order traits.

Newer data are also compatible with this conclusion about the critical importance of gene flow for determining patterns in second-order traits. Field and laboratory experiments on phenotypic plasticity identical to our previously published studies (Trexler and Travis 1990; Trexler et al. 1990) show that fish from populations in South Carolina, at the northern edge of the species' range, show norms of reaction to temperature and salinity variation that are markedly different from those seen in north Florida fish (from which they are separated by far more than the 200 km indicated by the spatial autocorrelation analyses as necessary for divergent allele frequencies), but that there is no heterogeneity in norms of reaction among fish from different local populations in South Carolina (Trexler and Travis, unpublished data).

The allozyme data may not indicate high levels of ongoing gene flow but may reflect low gene flow among recently separated, numerically large populations (Trexler 1988). If this scenario were true, local differentiation in body size would reflect a rapidly achieved, fine-scale adaptation that is made possible by relative isolation. Evidence directly contradicting this scenario comes from comparing the body-size distribution of male progeny reared in the laboratory from wild-caught, field-inseminated females with the distributions of male body size in fish from field collections that should represent the same generation at maturity in the field (fig. 2 of Trexler et al. 1990). Boat Ramp and Lighthouse Pond are separated by < 2 km of canals, closely adjacent physically, and they are very similar populations allozymically (Trexler 1988). Boat Ramp is a "large" population and Lighthouse Pond a "small" one with respect to male body size (Farr et al. 1986). Lab rearings from Boat Ramp females collected in the spring produce too many small males, whereas those from Lighthouse Pond produce too many large males, exactly the patterns to be expected if ongoing gene flow (via spring tidal movements) is countered by selection. It is also possible that the heterozygote deficiencies in adult fish alluded to above are a transient effect produced by high levels of reciprocal gene flow between populations differentiated at key loci.

Local variation in secondary sex characters could also be molded by the interplay of gene flow and sexual selection. Once the balance point for male body size is attained, continuing sexual selection could only engender a response in secondary sex characters (see Travis, in press, for a more detailed genetic argument). If males that appear larger and more vigorous because of higher dorsal fins and increased display rates can dominate smaller males and elicit more female cooperation, especially from larger females, then sexual selection can increase dorsal-fin size and display rates. This soft selection (Wade 1985) can work in the face of gene flow because it is for a primary trait (shape independent of size), not a secondary trait (a level of allometry between two traits). This hypothesis would explain the "countergradient" patterns of larger fins and higher display rates on "smaller" bodies. The presence of genetic variation for fin height and display rate for a given body size within a population (Travis et al., in press [a,b]) makes this a feasible hypothesis. This system may be an excellent example of the role of sexual selection *sensu stricto* in converting independent genetic variation for fin height and display rate within populations into concordant genetic variation among populations (O'Donald 1980, 1983; Arnold 1983; Kirkpatrick 1986).

Conclusions

The qualitative patterns of differentiation in primary and secondary traits exhibited by the array of sailfin molly populations are in accord with expectations derived from current models of the microevolutionary process. This agreement indicates that our current conceptual approaches offer a useful description of the microevolutionary process, despite the burden of their often untested assumptions (e.g., weak selection, local heterosis). Qualitative agreement is not really surprising; many theoretical results about broad-scale patterns are robust to deviations from many of their assumptions, especially in moderate to large populations (see Trexler 1988; also Travis and Mueller 1989).

From the point of view of "neglected factors," the most interesting result of these studies may be the indication that gene flow is a key determinant of the overall phenotypic patterns. Gene flow plays a critical role in studies of population structure, but it gets little attention as a contributor to phenotypic variance. In the molly system, it is clear that it represents little more than the elimination of immigrant genes from a local population in every generation. It is possible that many of the published observations of strong phenotypic selection (Endler 1986) represent just this phenomenon (e.g., Grant 1986).

This conclusion leads inevitably to the recommendation that any ecological genetic study of local variation ought to consider the larger spatial scale of adjacent populations and the extent to which interconnectedness among populations affects local processes (Via, chap. 4, this volume; Antonovics et al., chap. 8, this volume). Obviously just the observational studies of an ensemble of local molly populations helped diagnose the likely selective forces acting in single molly populations; this alone offers a cogent argument for working at larger spatial scales. However, it is clear in retrospect that the focus on a larger scale led to a more accurate picture of the local dynamics than would have been obtained otherwise.

These studies repeat the important lesson that similar patterns can arise from distinct processes. Even when the net selection pressure on a trait is in the same direction in different locations, different ecological causes may be responsible in the different locations. Small males are favored in north Florida as an indirect result of selection against long maturation times, whereas it appears that wading predation selects directly for small body size in the Everglades.

The recognition of such a difference in process is important for two reasons. First, any predictive understanding of selection requires knowledge of ecological causation; descriptions of selection without knowledge of ecological causation have little to offer anymore. Second, when distinct ecological factors produce the same net selective pressure on a character, the different ecological contexts of selection may also generate consequently different patterns of selection on other traits or different patterns of correlational selection (Phillips and Arnold 1989) on the original trait and other traits. Reynolds (in press) has shown in a simple model and experiment how such "context dependent selection" can select for different combinations of behavioral traits. Workers interested in whether selection alters genetic covariances among characters or produces divergent second-order patterns might profitably focus their empirical attention on just this type of situation.

From the point of view of "old lessons relearned," it seems obvious that a comprehensive study of population structure, life history, and behavior in one system provided more insight than each might have done separately. For example, the allozyme study, while useful in itself, provided a solid null hypothesis for how male body-size variation should be structured geographically if it merely followed gene-flow patterns. Some of the conclusions offered in the previous section draw convincing support only from the convergence of several disparate lines of evidence.

More critically, organisms are not as specialized as their students, and evolution does indeed act on the form of the whole organism. It should be apparent that selection has not molded variation in each trait of these animals independently of the other traits, and it has not done so independently of the population structure and zoogeographic history within which its actions have been embedded. The integration of knowledge from each of these avenues allows at least an outline of the microevolutionary process to be drawn. With more such outlines we will begin to understand more generally how the spatial scale of local adaptation is set and how evolution accommodates the diverse selection pressures on many individual traits of the same organism.

Acknowledgments

I have had the benefit of working with superb colleagues over many years. The ideas and work described here were developed in association with J. A. Farr and J. C. Trexler; who thought of what, and when, is lost to memory. An army of assistants has aided this project over the years; I am grateful to all of them, but the contributions of D. Buckheister, B. Fay, C. Johnson, S. Scheffer, and M. Trexler deserve special recognition. This work has been supported primarily by awards from the National Science Foundation and received additional support from the Florida Game and Fresh Water Fish Commission.

References

Antonovics, J., A. D. Bradshaw, and R. G. Turner. 1971. Heavy metal tolerance in plants. *Adv. Ecol. Res.* 7: 1–85.

Arnold, S. J. 1983. Sexual selection: The interface of theory and empiricism. In *Mate Choice*, ed. P. Bateson, pp. 67–107. Cambridge University Press, Cambridge, U.K.

Avise, J. C., J. C. Trexler, J. Travis, and W. L. Nelson. 1991. *Poecilia mexicana* is the recent female parent of the unisexual fish *P. formosa. Evolution* 45: 1530–1533.

Ayres, F. A., and S. J. Arnold. 1983. Behavioural variation in natural populations. IV. Mendelian models and heritability of a feeding response in the garter snake, *Thamnophis elegans. Heredity* 51: 405–413.

Baird, R. C. 1968. Aggressive behavior and social organization in *Mollienesia latipinna* (Le Sueur). *Texas J. Sci.* 20: 157–176.

Baird, R. C. 1974. Aspects of social behavior in *Poecilia latipinna* (Le Sueur). *Rev. Biol. Trop.* 21: 399–416.

Ball, R. M., S. Freeman, F. C. James, E. Bermingham, and J. C. Avise. 1988. Phylogeographic population structure of red-winged blackbirds assessed by mitochondrial DNA. *Proc. Natl. Acad. Sci. USA* 85: 1558–1562.

Ballinger, R. E. 1979. Intraspecific variation in demography and life history of the lizard, *Sceloporus jarrovi*, along an altitudinal gradient in Southeastern Arizona. *Ecology* 60: 901–909.

Balsano, J. S., E. J. Randle, K. Kucharski, E. M. Rasch, and P. J. Monaco. 1981. Reduction of competition between bisexual and unisexual females of *Poecilia* in northeastern Mexico. *Env. Biol. Fish.* 6: 39–48.

Bernays, E. A. 1986. Diet-induced head allometry among foliage-chewing insects and its importance for graminivores. *Science* 231: 495–497.

Berven, K. A. 1982. The genetic basis of altitudinal variation in the wood frog *Rana sylvatica*. I. An experimental analysis of life history traits. *Evolution* 36: 962–983.

Berven, K. A. 1987. The heritable basis of variation in larval developmental patterns within populations of the wood frog (*Rana sylvatica*). *Evolution* 41: 1088–1097.

Berven, K. A., D. E. Gill, and S. J. Smith-Gill. 1979. Countergradient selection in the green frog, *Rana clamitans. Evolution* 33: 609–623.

Borowsky, R. 1984. The evolutionary genetics of *Xiphophorus*. In *Evolutionary Genetics of Fishes*, ed. B. J. Turner, pp. 235–310. Plenum, New York.

Borowsky, R. L., and J. Diffley. 1981. Synchronized maturation and breeding in natural populations of *Xiphophorus variatus* (Poeciliidae). *Env. Biol. Fish.* 6: 49–58.

Bradshaw, W. E. 1986. Variable iteroparity as a life-history tactic in the pitcher-plant mosquito *Wyeomyia smithii. Evolution* 40: 471–478.

Breden, F., M. Scott, and E. Michel. 1987. Genetic differentiation for anti-predator behaviour in the Trinidad guppy, *Poecilia reticulata. Animal Behav.* 35: 618–620.

Brett, J. R. 1979. Environmental factors on growth. In *Fish Physiology*, ed. W. S. Hoar and D. J. Randall, vol. 8, pp. 599–675. Academic Press, New York.

Britton, R. H., and M. E. Moser. 1982. Size specific predation by herons and its effect on the sex ratio of natural populations of the mosquitofish *Gambusia affinis* Baird and Gerard. *Oecologia* 53: 146–151.

Bürger, R. 1986. Constraints for the evolution of functionally coupled characters: A non-linear analysis of a phenotypic model. *Evolution* 40: 182–193.

Caisse, M., and J. Antonovics. 1978. Evolution in closely adjacent plant populations. IX. Evolution of reproductive isolation in clinal populations. *Heredity* 40: 371–384.

Chappell, M. A., and L.R.G. Snyder. 1984. Biochemical and physiological correlates of deer mouse α-chain hemoglobin polymorphisms. *Proc. Natl. Acad. Sci. USA* 81: 5484–5488.

Charlesworth, B. 1980. *Evolution in Age-structured Populations*. Cambridge University Press, Cambridge, U.K.

Cockerham, C. C., and H. Tachida. 1988. Permanancy of response to selection for quantitative characters in the finite populations. *Proc. Natl. Acad. Sci. USA* 85: 1563–1565.

Conover, D. O. 1984. Adaptive significance of temperature-dependent sex determination in a fish. *Amer. Nat.* 123:297–313.

Conover, D. O., and S. W. Heins. 1987. Adaptation variation in environmental and genetic sex determination in a fish. *Nature* 326: 496–498.

Conover, D. O., and M. R. Ross. 1982. Patterns in seasonal abundance, growth, and biomass of the Atlantic silverside, *Menidia menidia*, in a New England estuary. *Estuaries* 5: 275–286.

Coyne, J. A., and E. Beecham. 1987. Heritability of two morphological characters within and among natural populations of *Drosophila melanogaster*. *Genetics* 117: 727–737.

Custer, T. W., and R. G. Osborn. 1978. Feeding-site description of three heron species near Beaufort, North Carolina. In *Research Report 7*, ed. A. Sprunt IV, J. C. Ogden, and S. Winckler, pp. 355–360. National Audubon Society, New York.

Eisen, E. J., J. P. Hanrahan, and J. E. Legates. 1973. Effects of population size and selection intensity on correlated responses to selection for postweaning gain in mice. *Genetics* 74: 157–170.

Endler, J. A. 1977. *Geographic Variation, Speciation, and Clines*. Princeton University Press, Princeton, N.J.

Endler, J. A. 1983. Natural and sexual selection on color patterns in poeciliid fishes. *Env. Biol. Fish.* 9: 173–190.

Endler, J. A. 1986. *Natural Selection in the Wild*. Princeton University Press, Princeton, N.J.

Evans, D. H. 1973. Sodium uptake by the sailfin molly, *Poecilia latipinna*: Kinetic analysis of a carrier system present in both fresh-water-acclimated and sea-water-acclimated individuals. *Comp. Biochem. Physiol.* 45A: 843–850.

Evans, D. H. 1975. The effects of various external cations and sodium transport inhibitors on sodium uptake by the sailfin molly, *Poecilia latipinna*, acclimated to sea water. *J. Comp. Physiol.* 96: 111–115.

Evans, D. H. 1984. The roles of gill permeability and transport mechanisms in euryhalinity. In *Fish Physiology*, ed. W. S. Hoar and D. J. Randall, vol. 10B, pp. 239–283. Academic Press, New York.

Farr, J. A., and J. Travis. 1986. Fertility advertisement by female sailfin mollies, *Poecilia latipinna* (Pisces: Poeciliidae). *Copeia* 1986: 467–472.

Farr, J. A., and J. Travis. 1989. The effect of ontogenetic experience on variation in growth, maturation, and sexual behavior in the sailfin molly, *Poecilia latipinna* (Pisces: Poeciliidae). *Env. Biol. Fish.* 26: 39–48.

Farr, J. A., J. Travis, and J. C. Trexler. 1986. Behavioural allometry and interdemic variation in sexual behaviour of the sailfin molly *Poecilia latipinna* (Pisces: Poeciliidae). *Animal Behav.* 34: 497–509.

Futuyma, D. J. 1983. Evolutionary interactions among herbivorous insects and plants. In *Coevolution*, ed. D. J. Futuyma and M. Slatkin, pp. 207–231. Sinauer, Sunderland, Mass.

Goodnight, C. J. 1988. Epistasis and the effect of founder events on the additive genetic variance. *Evolution* 42: 441–454.

Grant, P. R. 1986. *Ecology and Evolution of Darwin's Finches*. Princeton University Press, Princeton, N.J.

Grier, H. J. 1973. Reproduction in the teleost *Poecilia latipinna*, an ultrastructural and photoperiodic investigation. Ph.D. diss., University of South Florida, Tampa.

Gustafson, D. L. 1981. The influence of salinity on plasma osmolality and routine oxygen consumption in the sailfin molly, *Poecilia latipinna* (Lesueur), from a freshwater and an estuarine population. M.S. thesis, University of Florida, Gainesville.

Hirshfield, M. F., C. R. Feldmeth, and D. L. Stoltz. 1980. Genetic difference in physiological tolerances of Amargosa pupfish (*Cyprinodon nevadensis*) populations. *Science* 207: 999–1000.

Hubbs, C. L. 1942. Species and hybrids of *Mollienesia*. *The Aquarium* 10: 162–168.

Hubbs, C. 1964. Interactions between a bisexual fish species and its gynogenetic sexual parasite. *Bull. Texas Mem. Mus.* 8: 1–72.

James, F. C. 1970. Geographic size variation in birds and its relationship to climate. *Ecology* 51: 365–390.

James, F. C. 1983. Environmental component of morphological differentiation in birds. *Science* 221: 184–186.

Kallman, K. D. 1989. Genetic control of size at maturity in *Xiphophorus*. In *Ecology and Evolution of Livebearing Fishes* (Poeciliidae), ed. G. K. Meffe and F. F. Snelson, pp. 163–184. Prentice Hall, Englewood Cliffs, N.J.

Kilby, J. D. 1955. The fishes of two Gulf coastal marsh areas of Florida. *Tulane Studies in Zool.* 2: 175–247.

Kirkpatrick, M. 1986. The handicap mechanism of sexual selection does not work. *Amer. Nat.* 127: 222–240.

Koehn, R. K., A. J. Zero, and J. G. Hall. 1983. Enzyme polymorphism and natural selection. In *Evolution of Genes and Proteins*, ed. M. Nei and R. K. Koehn, pp. 115–136. Sinauer, Sunderland, Mass.

Kramer, D. L., D. Manley, and R. Bourgeois. 1983. The effect of respiratory mode and oxygen concentration on the risk of aerial predation in fishes. *Canadian J. Zool.* 61: 653–665.

Kushlan, J. A. 1978. Feeding ecology of wading birds. In *Research Report 7*, ed. A. Sprunt IV, J. C. Ogden, and S. Winckler, pp. 249–297. National Audubon Society, New York.

Lande, R. 1983. The response to selection on major and minor mutations affecting a metrical trait. *Heredity* 50: 47–65.

Laurie-Ahlberg, C. C., J. H. Williamson, B. J. Cochrane, A. N. Wilton, and F. I. Chasalow. 1981. Autosomal factors with correlated effects on the activities of the glucose 6-phosphate and 6-phosphogluconate dehydrogenases in *Drosophila melanogaster*. *Genetics* 99: 127–150.

Lee, D. S., C. R. Gilbert, C. H. Hocutt, R. E. Jenkins, D. E. McAllister, and J. R. Stauffer, Jr. 1980. *Atlas of North American Freshwater Fishes*. North Carolina State Museum of Natural History, Raleigh, N.C.

Lenski, R. E. 1988. Experimental studies of pleiotropy and epistasis in *Escherichia coli*. II. Compensation for maladaptive effects associated with resistance to virus T4. *Evolution* 42: 433–440.

Lloyd, M. 1967. Mean crowding. *J. Anim. Ecol.* 36: 1–30.

Lonsdale, D. J., and J. S. Levinton. 1985. Latitudinal differentiation in copepod growth: An adaptation to temperature. *Ecology* 66: 1397–1407.

Luckner, C. L. 1979. Morphological and behavioral polymorphism in *Poecilia latipinna* males (Pisces: Poeciliidae). Ph.D. diss., Louisiana State University, Baton Rouge.

Lynch, M. 1985. Spontaneous mutations for life-history characters in an obligate parthenogen. *Evolution* 39: 804–818.

MacDonald, J. F., G. K. Chambers, J. David, and F. J. Ayala. 1978. Adaptive response due to changes in gene regulation: A study with *Drosophila*. *Proc. Natl. Acad. Sci. USA* 24: 4562–4566.

Matsuo, Y., and T. Yamazaki. 1984. Genetic analysis of natural populations of *Drosophila melanogaster* in Japan. IV. Natural selection on the inducibility, but not on the structural genes, of amylase loci. *Genetics* 108: 879–896.

Mettler, L. E., T. G. Gregg, and H. E. Schaffer. 1988. *Population Genetics and Evolution*. 2d ed. Prentice Hall, Englewood Cliffs, N.J.

Monaco, P. J., E. M. Rasch, and J. S. Balsano. 1981. Sperm availability in naturally occurring bisexual-unisexual breeding complexes involving *Poecilia mexicana* and the gynogenetic teleost, *Poecilia formosa*. *Env. Biol. Fish.* 6: 159–166.

Moody, M. 1981. Polymorphism with selection and genotype-dependent migration. *J. Math. Biol.* 11: 245–267.

Nagahama, Y. 1983. The functional morphology of teleost gonads. In *Fish Physiology*, ed. W. S. Hoar, D. J. Randall, and E. M. Donaldson, vol. 9, pp. 223–275. Academic Press, New York.

Nagylaki, T. 1975. Conditions for the existence of clines. *Genetics* 80: 595–615.

Nagylaki, T., and B. Lucier. 1980. Numerical analysis of random drift in a cline. *Genetics* 94: 497–517.

Nevo, E., and S. Y. Yang. 1979. Genetic diversity and climatic determinants of tree frogs in Israel. *Oecologia* 41: 47–63.

O'Donald, P. 1980. *Genetic Models of Sexual Selection.* Cambridge University Press, Cambridge, U.K.

O'Donald, P. 1983. Sexual selection by female choice. In *Mate Choice,* ed. P. Bateson, pp. 53–66. Cambridge University Press, Cambridge, U.K.

Parzefall, J. 1969. Zur vergleichenden Ethologie verschiedener *Mollienesia*-Arten einschliesslich einer Höhlenform von *M. sphenops. Behaviour* 35: 1–37.

Parzefall, J. 1973. Attraction and sexual cycle of Poeciliidae. In *Genetics and Mutagenesis of Fish,* ed. J. H. Schröder, pp. 177–183. Springer-Verlag, Berlin.

Parzefall, J. 1979. Zur Genetik und biologischen Bedeutung des Aggressionsverhaltens von *Poecilia sphenops* (Pisces, Poeciliidae). *Z. Tierpsychol.* 50: 399–422.

Phillips, P., and S. J. Arnold. 1989. Visualizing multivariate selection. *Evolution* 43: 1209–1222.

Policansky, D. 1983. Size, age, and demography of metamorphosis and sexual maturation in fishes. *Amer. Zool.* 23: 57–63.

Powers, D. A., L. DiMichele, and A. R. Place. 1983. The use of enzyme kinetics to predict differences in cellular metabolism, developmental rate, and swimming performance between LDH-B genotypes of the fish, *Fundulus heteroclitus. Isozymes: Current Topics in Biol. and Med. Res.* 10: 147–170.

Powers, D. A., I. Ropson, D. C. Brown, R. Van Beneden, R. Cashon, L. I. Gonzalez-Villasenor, and J. A. DiMichele. 1986. Genetic variation in *Fundulus heteroclitus*: Geographic distribution. *Amer. Zool.* 26: 131–144.

Reynolds, J. D. In press. Should attractive individuals court more? Theory and a test. *Amer. Nat.*

Reznick, D. 1982. The impact of predation on life history evolution in Trinidadian guppies: Genetic basis of observed life history patterns. *Evolution* 36: 1236–1250.

Simanek, D. E. 1978. Population genetics and evolution in the *Poecilia formosa* complex (Pisces: Poeciliidae). Ph.D. diss., Yale University, New Haven, Conn.

Slatkin, M. 1973. Gene flow and selection in a cline. *Genetics* 75: 733–756.

Slatkin, M. 1975. Gene flow and selection in a two locus system. *Genetics* 81: 787–802.

Slatkin, M. 1978. Spatial patterns in the distributions of polygenic characters. *J. Theoret. Biol.* 70: 213–228.

Smith, M. H., M. C. Wooten, and K. Scribner. 1989. Genetic variation within and among *Gambusia* populations in the southeastern United States. In *Ecology and Evolution of Livebearing Fishes (Poeciliidae),* ed. G. K. Meffe and F. F. Snelson, Jr., pp. 235–257. Prentice Hall, Englewood Cliffs, N.J.

Snelson, F. F., Jr. 1982. Indeterminate growth in males of the sailfin molly, *Poecilia latipinna. Copeia* 1982: 296–304.

Snelson, F. F., Jr. 1985. Size and morphological variation in males of the sailfin molly, *Poecilia latipinna. Env. Biol. Fish.* 13: 35–47.

Snelson, F. F., Jr., J. D. Wetherington, and H. L. Large. 1986. The relationship between interbrood interval and yolk loading in a generalized poeciliid fish, *Poecilia latipinna. Copeia* 1986: 295–304.

Stanley, S. M. 1988. Fossils, macroevolution, and theoretical ecology. In *Perspec-*

tives in Ecological Theory, ed. J. Roughgarden, R. M. May, and S. A. Levin, pp. 125–147. Princeton University Press, Princeton, N.J.

Stearns, S. C. 1983. The evolution of life-history traits in mosquitofish since their introduction to Hawaii in 1905: Rates of evolution, heritabilities, and developmental plasticity. *Amer. Zool.* 23: 65–75.

Stearns, S. C. 1984. Heritability estimates for age and length at maturity in two populations of mosquitofish that shared ancestors in 1905. *Evolution* 38: 368–375.

Stearns, S. C., and J. C. Koella. 1986. The evolution of phenotypic plasticity in life-history traits: Predictions of reaction norms for age and size at maturity. *Evolution* 40: 893–913.

Sumner, I. T., J. Travis, and C. D. Johnson. In press. Methods of female fertility advertisement and variation among males in responsiveness in the sailfin molly (*Poecilia latipinna*). *Copeia.*

Tachida, H., and C. C. Cockerham. 1987. Quantitative genetic variation in an ecological setting. *Theoret. Pop. Biol.* 32: 393–429.

Travis, J. 1989. Ecological genetics of life-history traits in poeciliid fishes. In *Ecology and Evolution of Livebearing Fishes (Poeciliidae)*, ed. G. K. Meffe and F. F. Snelson, Jr., pp. 185–200. Prentice Hall, Englewood Cliffs, N.J.

Travis, J. In press. Behavioral allometry: The genetics of correlated trait variation. In *Quantitative Genetic Approaches to Animal Behavior*, ed. C. M. Boake. University of Chicago Press, Chicago.

Travis, J., and Mueller, L. D. 1989. Blending ecology and genetics: Progress toward a unified population biology. In *Perspectives in Ecological Theory*, ed. J. Roughgarden, R. M. May, and S. A. Levin, pp. 101–124. Princeton University Press, Princeton, N.J.

Travis, J., and J. C. Trexler. 1987. Regional variation in habitat requirements of the sailfin molly, with special reference to the Florida Keys. *Nongame Wildlife Technical Report Number 3*, Florida Game and Fresh Water Fish Commission, Tallahassee.

Travis, J., and B. D. Woodward. 1989. Social context and courtship flexibility in male sailfin mollies, *Poecilia latipinna* (Pisces: Poeciliidae). *Animal Behav.* 38: 1001–1011.

Travis, J., J. A. Farr, M. McManus, and J. C. Trexler. 1989. Environmental effects on adult growth patterns in the male sailfin molly (*Poecilia latipinna*). *Env. Biol. Fish.* 26: 119–127.

Travis, J., J. C. Trexler, and M. M. Mulvey. 1990. Multiple paternity and its correlates in female *Poecilia latipinna* (Pisces, Poeciliidae). *Copeia* 1990: 722–729.

Travis, J., J. A. Farr, and J. C. Trexler. In press (a). Body-size variation in the sailfin molly, *Poecilia latipinna* (Pisces: Poeciliidae): II. Genetics of male behaviors and secondary sex characters. *J. Evol. Biol.*

Travis, J., J. C. Trexler, and J. A. Farr. In press (b). Body-size variation in the sailfin molly, *Poecilia latipinna* (Pisces: Poeciliidae): I. Sex-limited genetic variation for size and age of maturation. *J. Evol. Biol.*

Trexler, J. C. 1985. Variation in the extent of viviparity in the sailfin molly, *Poecilia latipinna*. *Copeia* 1985: 999–1004.

Trexler, J. C. 1986. Geographic variation in size in the sailfin molly, *Poecilia latipinna*. Ph.D. diss., Florida State University, Tallahassee.

Trexler, J. C. 1988. Hierarchical organization of genetic variation in the sailfin molly, *Poecilia latipinna* (Pisces, Poeciliidae). *Evolution* 42: 1006–1017.

Trexler, J. C. 1989. Phenotypic plasticity in poeciliid life histories. In *Ecology and Evolution of Livebearing Fishes (Poeciliidae)*, ed. G. K. Meffe and F. F Snelson, Jr., pp. 201–214. Prentice Hall, Englewood Cliffs, N.J.

Trexler, J. C., and J. Travis. 1990. Phenotypic plasticity in the sailfin molly, *Poecilia latipinna* (Pisces: Poeciliidae). I. Field experiment. *Evolution* 44: 143–156.

Trexler, J. C., R. C. Tempe, and J. Travis. In press. Size-selective predation of sailfin mollies by two species of heron. *Oikos*.

Trexler, J. C., J. Travis, and M. McManus. 1992. Effects of habitat and body size on mortality rates of *Poecilia latipinna*. *Ecology*. 73: 2224–2236.

Trexler, J. C., J. Travis, and M. Trexler. 1990. Phenotypic plasticity in the sailfin molly, *Poecilia latipinna* (Pisces: Poeciliidae). II. Laboratory experiment. *Evolution* 44: 157–167.

Via, S., and R. Lande. 1985. Genotype-environment interaction and the evolution of phenotypic plasticity. *Evolution* 39: 505–522.

Wade, M. J. 1985. Soft selection, hard selection, kin selection, and group selection. *Amer. Nat.* 125: 61–73.

Williams, G. C., and R. K. Koehn. 1984. Population genetics of North American catadromous eels (*Anguilla*). In *Evolutionary Genetics of Fishes*, ed. B. J. Turner, pp. 529–560. Plenum, New York.

Wilton, A. N., C. C. Laurie-Ahlberg, T. H. Emigh, and J. W. Curtsinger. 1982. Naturally occurring enzyme activity variation in *Drosophila melanogaster*. II. Relationship among enzymes. *Genetics* 102: 207–221.

Woodhead, A. D., and N. Armstrong. 1985. Aspects of the mating behavior of male mollies (*Poecilia* spp.). *J. Fish. Biol.* 27: 593–601.

Zeng, Z.-B. 1988. Long-term correlated response, interpopulation covariation, and interspecific allometry. *Evolution* 42: 363–374.

Zimmerer, E. J. 1983. Effect of salinity on the size-hierarchy effect in *Poecilia latipinna*, *P. reticulata*, and *Gambusia affinis*. *Copeia* 1983: 243–245.

Zimmerer, E. J., and K. D. Kallman. 1988. The inheritance of vertical barring (aggression and appeasement signals) on the pygmy swordtail, *Xiphophorus nigrensis* (Poeciliidae, Teleostei). *Copeia* 1988: 299–307.

Index